PESTICIDE BIOTRANSFORMATION AND DISPOSITION

PESTICIDE BIOTRANSFORMATION AND DISPOSITION

A further development from HAYES' HANDBOOK OF PESTICIDE TOXICOLOGY, Third Edition.
(Robert Krieger, Editor)

Editor

ERNEST HODGSON

Amsterdam • Boston • Heidelberg • London • New York • Oxford
Paris • San Diego • San Francisco • Singapore • Sydney • Tokyo
Academic Press is an imprint of Elsevier

Academic Press is an imprint of Elsevier
32 Jamestown Road, London NW1 7BY, UK
225 Wyman Street, Waltham, MA 02451, USA
525 B Street, Suite 1800, San Diego, CA 92101-4495, USA

First edition 2012

Notice
No responsibility is assumed by the publisher for any injury and/or damage to persons or property as a matter of products liability, negligence or otherwise, or from any use or operation of any methods, products, instructions or ideas contained in the material herein. Because of rapid advances in the medical sciences, in particular, independent verification of diagnoses and drug dosages should be made

British Library Cataloguing-in-Publication Data
A catalogue record for this book is available from the British Library

Library of Congress Cataloging-in-Publication Data
A catalog record for this book is available from the Library of Congress

ISBN : 978-0-12-385481-0

For information on all Academic Press publications
visit our website at elsevierdirect.com

Typeset by MPS Limited, a Macmillan Company, Chennai, India
www.macmillansolutions.com

Printed and bound in United States of America

10 11 12 13 14 15 10 9 8 7 6 5 4 3 2 1

CONTENTS

DEDICATION FROM *HAYES' HANDBOOK OF PESTICIDE TOXICOLOGY*, THIRD EDITION

Wayland Jackson "Jack" Hayes, Jr. made enduring contributions to pesticide science. *Hayes' Handbook of Pesticide Toxicology*, third edition, carries his name to recognize his profound commitment to "improve the knowledge of toxicology, in general, the epidemiology of pesticide poisoning, and the medical management of cases." He wrote and spoke often of the importance of the first principles of toxicology as Chief Toxicologist at Centers for Disease Control, Atlanta, Georgia, and later as Professor of Toxicology, School of Medicine, Vanderbilt University, Nashville, Tennessee.

Hayes contributed his first volume to the toxicological literature as the *Clinical Handbook on Economic Poisons* (1963), replacing "Clinical Memoranda on Economic Poisons" first issued in March 1950 as separate releases on several new insecticides. The booklet described the diagnosis and treatment of persons who may have had extensive or intensive exposure to economic poisons. It was prepared primarily for the guidance of physicians and other public health professionals. The 1963 booklet concerned the use of organophosphorus insecticides and acute toxicities associated with pesticides such as "arsenic, thallium, phosphorous, and kerosene" because they were "leading causes of deaths associated with pesticides". Hayes acknowledged the great potential value of the materials used as pesticides and urged the careful collection of clinical data and related information concerning poisoning, a theme that became much clearer in the expanded *Toxicology of Pesticides* (1975). *Toxicology of Pesticides* and his works that followed gave attention to "those materials that are manufactured in large amounts, that are known to have caused poisoning relatively frequently, or that are of special interest for some other reason". The subjects of clinical studies included: (1) persons with "heavy occupational exposure"—including malaria control spray operators, farmers, orchardists, spray pilots, and pest control operators; (2) volunteers who take part in strictly controlled experimental investigations; and (3) patients who are sick from accidental over-exposure to pesticides. In the preface to his next major work and the first edition in the present series, he called attention to the need for basic toxicology education. *Pesticides Studied in Man* (1982) and *The Handbook of Pesticide Toxicology* represent his commitment to the collection and dissemination of critical research and clinical experience in Hayes' career as a leader in pesticide science.

Widespread use of the *Clinical Handbook on Economic Poisons* and active participation in public debate concerning pesticide use encouraged Hayes to write of the general importance of the principles of toxicology. In *Toxicology of Pesticides* (1975) and his subsequent books he retained the strong clinical content but offered much expanded coverage of principles of toxicology, the conditions of exposure, the effects on human health, problems of diagnosis and treatment, the means to prevent injury, and even brief outlines on the impact of pesticides on domestic animals and wildlife.

In the public arena, Hayes spoke out on an expanding role of toxicology to address issues of public and environmental health related to pesticide use that became critical during the 1960s and 1970s following publication of Rachel Carson's polemic *Silent Spring* (1962). Concerning the resulting intense public debate about pesticides, Hayes wrote in the Preface to *Toxicology of Pesticides*:

> *"The pesticide problem is not merely one concerning the chemical industry and professional farmers, foresters, and applicators, or one concerning only those who wish to protect wildlife, or those responsible for control of malaria and other vector-borne diseases of man and his livestock. Rather, the pesticide problem concerns every person who wants food at a reasonable price and who wants his home free from vermin. The problem can be solved only on the basis of sound toxicological principles. Knowledge of these principles permits agreement and a cooperative approach on the part of persons professionally responsible for protection of our food, our health, and our wildlife, respectively. Ignorance of these principles limits some other persons to a partisan approach that may be dangerous to the common good."*

In dedicating *Toxicology of Pesticides* to Paracelsus, Hayes sought to bring attention to the "decisive importance of dosage" in determining the effect of exposure. He urged recognition of "tolerated doses" as well as information on doses or blood levels that have produced harm. He clearly viewed modern toxicology as a predictive, interdisciplinary science with great capacity to contribute to chemical safety evaluation.

His *Pesticides Studied in Man* (1982) assumed the reader's mastery of the basic principles of toxicology and offered more in-depth coverage of those pesticides with direct information concerning their effects in humans. The information came from reports of poisoning, from observation of workers or volunteers, or from persons who received certain compounds as drugs. Sections were organized in three parts. The first gave a concise summary of the chemistry and use of the pesticide. The second part concerned the fate and basic animal toxicity data that contributed to determining important dose-response relationships. The third section reported the human experience with the pesticide. The present edition of *Hayes' Handbook of Pesticide Toxicology* applies this basic scheme more loosely in the description of the toxicology of agents.

As Professor of Biochemistry, School of Medicine, Vanderbilt University, Hayes teamed with his colleague Edward R. Laws, Jr., Department of Neurological Surgery, George Washington School of Medicine, Washington, D. C. to edit the first edition of the *Handbook of Pesticide Toxicology*. It was published by Academic Press in three volumes and updated and revised both *Toxicology of Pesticides* and *Pesticides Studied in Man*. The Preface again champions the potential role of toxicology in the resolution of controversy regarding pesticide use and reiterates the importance of the study of dose-response relationships in diagnosis of poisoning. The book follows familiar organization, including exposition of principles of toxicology and sections featuring the chemistry and uses of pesticides, biochemistry and experimental toxicology, and description of the human experience with pesticides.

Hayes' admonition to physicians to collect quantitative information on the effects of different dosages is consistent with his high regard for the fullest possible data concerning the human experience with pesticides. Throughout his career Hayes shaped a vision of modern toxicology as an important means to achieve rational use of chemicals in the environment, much in the spirit of Paracelsus who wrote, "… whenever I went I eagerly and diligently investigated and sought after the tested and reliable arts of medicine. I went not only to the doctors, but also to barbers, bathkeepers, learned physicians, women, and magicians who pursue the art of healing".

Wayland Hayes was born in Charlottesville, Virginia, on April 29, 1917. He graduated in 1938 from the University of Virginia, received an M. A. degree and a Ph.D. from the University of Wisconsin where he specialized in zoology and physiological chemistry. He returned to the University of Virginia where he received his M.D. in 1946. He interned in the Public Health Service Hospital in Staten Island, New York, and entered the regular corps of the service from 1948 to 1968. He became Chief Toxicologist of the Pesticides Program of the Centers for Disease Control in Savannah and Atlanta, Georgia. Hayes joined Vanderbilt University as Professor of Biochemistry, School of Medicine, in 1968 becoming *emeritus* in 1982 but remaining active in university affairs until 1991. He died January 4, 1993. His wife, Barnita Donkle Hayes, of 50 years and a son, Wayland J. Hayes III, and four daughters, Marie Royce Hayes, Maryetta Hayes Hacskaylo, Lula Turner McCoy and Roche Del Moser; and 10 grandchildren, survived him. In his family and community, he was revered as a parent, gardener, artist, philosopher and humorist.

Hayes had a full professional life of national and international service. He was a consultant on the toxicology of pesticides to the World Health Organization, the Pan American Sanitary Bureau, the American Medical Association, the U. S. Department of Agriculture/Environmental Protection Agency, the American Conference of Governmental Industrial Hygienists and the National Academy of Sciences-National Research Council. He served on numerous governmental committees and editorial boards. He was a charter member of the Society of Toxicology in 1961 and served as its eleventh president 1971–72. As president of the Society, he staunchly defended the integrity of toxicologists in regulatory affairs (Science 174: 545–546, 1971) and launched criticism of the USEPA's dismissal of the recommendation of its own Scientific Advisory Committee in response to "external pressure". As president, Hayes made a strong plea for the inclusion of toxicology in textbooks of biology, zoology, hygiene, and general science (Toxicology and Applied Pharmacology 19, i–ii, 1971). Both subjects are topical today. Other society memberships included the American Society of Pharmacology and Experimental Therapeutics and the American Society of Tropical Medicine and Hygiene. He became a Diplomat of The Academy of Toxicological Sciences in 1989.

Wayland Hayes was a sought after expert witness, particularly in cases involving pesticides. His commanding and distinguished presence, his southern accent and gracious manner coupled with his encyclopedic knowledge rarely failed to win the case. However, there was one case in Wisconsin where he was unable to convince the jury that DDT was not a potent poison. Finally, he walked over to the evidence table, picked up the bottle of DDT and ingested a teaspoon of the evidence. When asked about how that worked out, he replied, "Well I may have walked a little funny, but we won the case".

Hayes clearly recognized the difficulties associated with collecting meaningful dosage-response information. He suggested that failure to collect such valuable data might result from lack of recognition of its importance in diagnostics. He closed on a theme that has shaped his career and that remains central to the spirit and content of the current volumes now dedicated to his life and career saying, "Clinicians who attend patients poisoned by a pesticide or by any other material are urged to be alert to the possibility of getting new information on dosage."

Robert I. Krieger, Ph.D.
John W. Doull, M.D., Ph.D.

Although this monograph is derived from the 3rd edition of the *Handbook of Pesticide Toxicology* (R. Krieger, editor, Elsevier, 2010) it is more than a reorganization of a set of related chapters. Of the 10 chapters, 4 are new. All of the remaining 6 chapters have been revised and updated to a greater or lesser extent. Those that provide essential background are not dramatically different from the corresponding chapter in the *Handbook*, while those that are intended to be a source of specific detailed information include new material.

The result is a monograph focused on a particular aspect of pesticide toxicology, an aspect that continues to develop and is of considerable importance, both fundamental and applied. Information on pesticide biotransformation is needed for human health risk assessment, the data from investigations on surrogate animals to aid in extrapolation to humans, while that on human biotransformation is of unique value in the assessment of human variation and in the definition of population subgroups and individuals at increased risk. Since pesticides continue to be regulated as single chemicals but are, more often than not, used as mixtures, knowledge of pesticide metabolism contributes to our emerging understanding of the problem of pesticide interactions in pesticide mixtures as used in the field.

Many thanks to Elsevier for adopting this means of extending the use of the *Handbook* in a practical, portable, and usable format and to their always helpful and willing aid and assistance in bringing the project to fruition.

Ernest Hodgson
June 2011
Raleigh, North Carolina

The Third edition of the *Handbook* is renamed the *Hayes' Handbook of Pesticide Toxicology* and dedicated to the memory of Wayland J. Hayes, Jr., whose major contributions to pesticide science are chronicled, in part, in the Dedication. The cover design includes the whimsical 3-segmented, 6-legged doodle by Hayes used on the First and Second editions of the Handbook.

This edition of *Hayes' Handbook of Pesticide Toxicology* includes primarily new and revised chapters concerning fundamentals of the past and new insights gained from more recent research in pesticide science. More complete exposition of the concepts which have guided preparation of these volumes is contained in the Preface to the Second edition included herein.

Pests and organisms that would devour our residences, personal property and food supply remain ever-present competitors in human environments. In response, pesticides delivered in developed nations with increasing precision and regulation represent a chemical technology that is refined, extensively used and studied in detail. Chemical exposures, particularly those related to the economic class *pesticide*, are an analytical reality that remains problematic for many persons in spite of overwhelming environmental monitoring which reveals that exposures occur at levels benign to health. The Handbook is expected to contribute to clarification, and even resolution, of some imperfections or limitations in available knowledge.

Numerous experts, more than 200 in all, have contributed their time and expertise to the Third edition. Their contributions are particularly noteworthy and appreciated in continued times of economic uncertainty, emerging. Regulatory priorities, and considerable instability in private and public institutions as priorities and programs take new forms. The authors have provided in-depth review and exposition of the particular topics that are included in this edition. References will allow interested readers to pursue topics of interest.

Each of the Associate Editors, including John Doull, Joop van Hemmen (deceased), Ernest Hodgson, Howard Maibach, Lawrence Reiter, Leonard Ritter, John Ross, and William Slikker is acknowledged and thanked for his important and particular contributions to the development and production of the *Hayes' Handbook of Pesticide Toxicology*. These volumes represent the tireless dedication and exemplary service of Helen Vega, Administrative Assistant in the Personal Chemical Exposure Program here at Riverside and Editorial Assistant for the *Hayes' Handbook of Pesticide Toxicology*. We are both grateful to Kirsten Chrisman, Rebecca Garay, April Graham, and Caroline Jones of Elsevier who effectively moved the author's copy to text.

Robert I. Krieger
University of California, Riverside

Dr. Ronald Baynes, Associate Professor of Pharmacology in the College of Veterinary Medicine at North Carolina State University. Dr. Baynes received his B.Sc. (with Honors) from the University of the West Indies (Cave Hill Campus), his D.V.M. (with Honors) from Tuskegee University, his M.S. from the University of Georgia, and his Ph.D. from North Carolina State University. His research is focused on using quantitative structure-activity relationship modeling approaches to understanding the physicochemical factors influencing dermal absorption of pesticides and formulation additives that cause occupational irritant dermatitis.

Dr. Kelly J. Dix, the author of Chapter 24 in the 2nd edition of the *Handbook of Pesticide Toxicology,* is no longer involved in toxicological pursuits or with the *Handbook.* However, her previous chapter was the progenitor of Chapter 39 in the *Handbook* and, as a result, contributed to it and to Chapter 6 of the current volume. Her contribution is much appreciated.

Dr. Ernest Hodgson is active as a Distinguished Professor Emeritus in the Department of Environmental and Molecular Toxicology at North Carolina State University and a member of the North Carolina Agromedicine Institute, a three-university consortium located at East Carolina University. He also serves as the Executive Director of the Foundation for Toxicology and Agromedicine. Dr. Hodgson was awarded his B.Sc. with Honors by King's College of the University of Durham (UK) (now the University of Newcastle) and his Ph.D. by Oregon State University. He has been interested in the metabolism of pesticides for many years and, more recently, has focused on the human metabolism of pesticides. Dr. Hodgson is also known for the publication of two widely accepted textbooks of toxicology. Dr. Hodgson is the Society of Toxicology 2012 Distinguished Toxicology Scholar awardee.

Dr. Chris Hofelt, DABT is an Assistant Professor in the Department of Environmental and Molecular Toxicology, as well as the undergraduate program coordinator. Dr. Hofelt holds a Ph.D. in Toxicology from North Carolina State University and board certification in General Toxicology from the American Board of Toxicology. He has worked in the environmental industry for over 19 years as a chemist, toxicologist, researcher, environmental regulator, risk assessor, and educator. Dr. Hofelt has been with North Carolina State as a faculty member since 2002 and has worked at the forefront of distance education with the university.

Dr. Jim E. Riviere is the Burroughs Wellcome Fund Distinguished Professor of Pharmacology and Director of the Center for Chemical Toxicology Research and Pharmacokinetics, College of Veterinary Medicine, North Carolina State University in Raleigh, North Carolina. Dr. Riviere received his B.S. (summa cum laude) and M.S. degrees from Boston College, his D.V.M. and Ph.D. in pharmacology, as well as a D.Sc. (with Honors) from Purdue University. He is an elected member of the Institute of Medicine of the National Academies, serves on its Food and Nutrition Board, and is a Fellow of the Academy of Toxicological Sciences. His current research interests relate to the development of animal models; applying biomathematics to problems in toxicology, including the risk assessment of chemical mixtures, pharmacokinetics, nanomaterials, and absorption of drugs and chemicals across skin; and the food safety and pharmacokinetics of tissue residues in food-producing animals.

CHAPTER 1

Introduction to Pesticide Biotransformation and Disposition

Ernest Hodgson
North Carolina State University, Raleigh, NC, USA

Outline

INTRODUCTION

It should be emphasized that, although pesticides and their use have many positive attributes, they are toxicants and, in terms of their interactions with living organisms, are xenobiotics, and are processed in the same way as other xenobiotics such as clinical drugs and industrial chemicals.

It should also be emphasized that their toxicity is not due to a single defining molecular event or interaction, but rather a cascade of events beginning with exposure and culminating with the expression of one or more toxic endpoints. This cascade (Figure 1.1) includes adsorption, distribution, metabolism (both detoxication and activation), distribution of metabolites, interaction with cellular macromolecules (such as RNA, DNA, and proteins), repair, and excretion. The processes involved may be reversible to a greater or lesser extent, they may include alternative pathways, and they may be modified by chemical and physiological interactions. Thus, exposure to a toxicant does not inevitably lead to a toxic endpoint; metabolism, excretion, or repair may render the original exposure without effect (Hodgson, 2010a). Finally, these processes and the genes, enzymes, transporters, receptors, etc., involved are all subject to considerable variation with cell type, organ, individual, species, and strain.

The aspects covered in this volume include adsorption, distribution, biotransformation (metabolism), and excretion and are collectively known as disposition. Biotransformation (metabolism), a more specialized term and a subdivision of disposition, of a xenobiotic is the total of all of the chemical transformations of that xenobiotic taking place in a living organism. In the case of xenobiotics (including pesticides), the use of the term disposition is often preferred to metabolism, since the latter is most often used to describe the total of

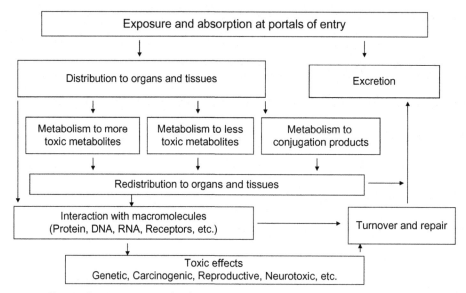

Figure 1.1 Chemical toxicity: a cascade of events.

all chemical reactions of normal body constituents. However, the two terms can usually be used as synonyms without confusion.

RELEVANCE OF BIOTRANSFORMATION AND DISPOSITION STUDIES

Studies of pesticide disposition, particularly biotransformation, are critical to the understanding of the toxic mode of action in both target and nontarget organisms. Biotransformation may result in the formation of less toxic (detoxication) and/or more toxic (activation) products, while the various other processes shown in Figure 1.1 may determine the balance between toxic and nontoxic events. While disposition in non-target species, including humans, is the primary focus of this volume, studies in target species facilitate the development of more effective, safer pesticides.

Most important, disposition studies facilitate risk analysis. They make possible physiologically based pharmacokinetic studies, since not only is knowledge of the uptake, distribution, and rate of excretion of the parent chemical necessary, but also that of the distribution and excretion of its metabolites. Mechanism of action at the molecular level cannot be defined unless all active metabolites have been identified and their interactions at the site of action determined. Quantitative structure activity relationship studies, important for the prediction of both effectiveness and toxicity, likewise depend on detailed knowledge of metabolites and their formation.

Since risk analysis of pesticides relies heavily on studies of single chemicals in surrogate animals, studies of pesticide metabolism in humans assume particular importance. Given the ready availability of hepatocytes, cell fractions, cell lines, and recombinant enzymes, all derived from humans, ethical human studies have been relatively easy to conduct for the past decade.

The surrogate animals used in metabolism studies, generally rodents, are highly inbred, while the human population is outbred and pesticides are, more often than not, used in mixtures or in temporal proximity so close as to have the same implications for risk analysis as mixtures. Thus studies in humans are essential if variation is part of the risk assessment paradigm and if subpopulations and individuals at increased risk are to be identified. They are also important in defining interactions between pesticides in mixtures and between pesticides and endogenous metabolites that may impact human health. Moreover, if surrogate animals are to be used, some studies in humans may indicate which experimental animal is the best surrogate for humans for studies of a particular pesticide or mixture of pesticides.

Given the emerging changes in risk assessment (National Research Council, 2007; Hodgson, 2010b; Kullman et al., 2010) that rely heavily on human cell lines and the techniques of genomics, proteomics, metabolomics, and informatics, the nature of human studies will doubtless change but their importance will increase.

REFERENCES

Committee on Toxicity Testing and Assessment of Environmental Agents, National Research Council. (2007). *Toxicity testing in the 21st century: A vision and a strategy.* Washington, DC: National Academies Press.

Hodgson, E. (2010a). Introduction to toxicology. In E. Hodgson (Ed.), *A textbook of modern toxicology* (4th ed.). Hoboken, NJ: John Wiley & Sons (Chap. 1).

Hodgson, E. (2010b). Future considerations. In E. Hodgson (Ed.), *A textbook of modern toxicology* (4th ed.). Hoboken, NJ: John Wiley & Sons (Chap. 29).

Kullman, S. W., Mattingly, C. J., Meyer, J. N., & Whitehead, A. (2010). Perspectives on informatics in toxicology. In E. Hodgson (Ed.), *A textbook of modern toxicology* (4th ed.). Hoboken, NJ: John Wiley & Sons (Chap. 28).

CHAPTER 2

Summary of Methods Used in the Study of Pesticide Biotransformation and Disposition

Chris Hofelt
North Carolina State University, Raleigh, NC, USA

Outline

INTRODUCTION

Food security and public health contribute to the foundations of a prosperous civil society. Pest management in general, and pesticides in particular, are of paramount importance in contributing to this foundation. Rapidly emerging technologies have contributed to the study of pesticide toxicology, allowing for the development of safer

Pesticide Biotransformation and Disposition
DOI: 10.1016/B978-0-12-385481-0.00002-2

and more effective pest management strategies. Although new techniques and instruments continue to enter the commercial market, the basic analytical process has not changed. One must:

- Define the research goal(s)
- Identify appropriate techniques and methods
- Develop a sampling scheme to obtain representative samples
- Isolate the compound(s) of interest
- Remove potential interfering components
- Quantify and evaluate the data in relation to the original research goals

Based on the data generated, many options are available. For example, was the sampling scheme complete? Would further refinement of the analytical procedure be required? Should other sample types be analyzed? Thus it is obvious that within these general categories, particular methods vary considerably depending on the chemical characteristics of the toxicant. This chapter is concerned with the sampling, isolation, separation, and measurement of chemicals, which are needed to study the biotransformation and disposition of pesticides. It is intended to give the reader a brief summary of some of the methods used in the elucidation of the disposition and metabolism of pesticides. Because an exhaustive treatment of this topic would require several volumes, the topics will be covered in such a way as to give the reader an overview and a "nodding acquaintanceship" with the topic. The reader is directed to the cited references for a more thorough treatment of individual topics.

ANALYTICAL METHODS FOR PESTICIDES AND PESTICIDE METABOLITES

Elucidation of the often complex mechanisms and pathways involved in pesticide biotransformation requires well-thought-out and well-controlled experiments. One of the cornerstones of these studies is good analytical chemistry methods that will produce robust and reliable data on the parent compound, metabolites, and (potentially) changes in gene and protein expression. This starts with an appropriate sampling technique, including not just the collection of the sample but an appropriate handling and storage process as well. The samples are then cleaned up (if necessary) and enriched so that they may be identified. There are a wide variety of techniques available; however, the use of mass spectrometry (MS) coupled with either gas chromatography (GC) or liquid chromatography (LC) is among the most prevalent. In this section we discuss these methodologies as well as tools used for toxicokinetics.

Sampling

Even with the most sophisticated analytical equipment available, the resulting data are only as representative as the samples from which the results are derived. The vast majority of errors in analytical measurements are introduced during the sample collection, preservation, and storage. Therefore a great deal of effort should go into the planning of this phase of the study and care must be taken to ensure that the resulting data meet the objectives of the study. Often special attention to sampling procedures is necessary. Ultimately, the initial hypothesis and goals of the study will determine what will be the most appropriate sampling schemes and analytical methods. In its simplest form, the goal in developing a sampling scheme is to come up with a representative sample, or representative samples. What this means is a sample that truly reflects the composition of the matrix to be analyzed within the context of the study aims.

In the context of this discussion, the samples of interest will tend to be biological. They may include whole blood, plasma, urine, bile/feces, cell culture media, saliva, adipose and organ tissue, or plant material. Each of these sample media has unique characteristics that could affect the choice of method, but there are a few constants that should be kept in mind. First, and foremost, any use of human tissue samples should have all appropriate informed consent and IRB approvals in place at the outset. With that said, there are some common considerations regardless of the matrix. Sample collection apparatus should be clean and of an appropriate material. For instance, highly hydrophobic compounds could adhere to polyethylene or silicone parts in a collection syringe and reduce the amount of compound available for analysis. Sample storage containers should likewise follow this rule. In general, glass containers with Teflon-lined lids are the most common. If the compound is photolabile, then the containers should be amber or otherwise protected from light. If the samples are not going to be analyzed immediately, then they should be stored at -20 to $-80°C$.

Blood
In human volunteers, this type of collection is relatively straightforward; in rodents, however, it is more problematic. Blood or plasma can be collected from laboratory animals through a variety of mechanisms. In smaller rodents it is especially important that the frequency of collection is not such that it significantly alters the total blood volume of the organism. In rats, 100- to 200-µl blood samples can be collected from the tail vein. In addition, a more complex method is to anesthetize the animal and insert a cannula into the external jugular vein. The orbital sinus in mice can be used to collect blood but only under terminal anesthesia, whereas in rabbits, the terminal ear vein may be used as a collection site without anesthesia (Hayes, 2008).

Urine and Bile
As with blood, collection of urine or fecal samples is not overly problematic with human volunteers. In rodents, urine and fecal samples can be collected by placing the animal in a metabolism cage. In this type of apparatus, the urine and feces are collected

via a stainless steel funnel with a steep-angled surface and which is covered with a special nonwetting material. The cone is equipped with an apparatus to separate the urine and feces into two different collection containers. The cages are also equipped with special food and water containers to prevent spillage into the sample collection area. For more on the collection of urine from laboratory animals, the reader is referred to Kurien et al. (2004) or Hayes (2008). Bile, on the other hand, can be collected via a cannula inserted into the bile duct of an anesthetized animal. However, not allowing the bile salts to circulate can lead to changes in the physiology of the animal.

Tissue

The collection of tissue can range from collecting small biopsies of muscle or adipose tissue to harvesting entire organs from necropsied animals. If one is investigating the biotransformation of pesticides in plants, it may include the collection of plant material such as leaves or stems. In all of these cases, the procedure is relatively uncomplicated and the biggest concern will tend to be the mass of sample available for collection, as this will affect the sensitivity of the analysis.

Other Methods

Although this is not an exhaustive list, there are a few other sampling methodologies that are worthy of note. The first is the use of microdialysis. The principle of dialysis is that compounds will partition through a semipermeable membrane, driven by the concentration gradient across the membrane. Although there are various designs, this sampling method generally uses a small concentric probe of dialysis tubing with a membrane at the tip. Perfusate can be cycled through the probe at a constant flow rate and collected in a vial (Plock and Kloft, 2005). The advantages of this type of sampling are that samples can be collected at the site of interest and the samples are free of protein that can confound analysis. In addition, samples can be collected over a period of time to determine toxicokinetic profiles without concern for blood loss in the animal. Some of the drawbacks of this technique include the potential for tissue trauma at the site of the probe insertion as well as the potential for bacterial contamination. In addition, the size of the sample is relatively small and therefore the analytical technique chosen has to have sufficient sensitivity to detect the analytes at very low concentrations. One such apparatus uses a microdialysis pump that directly injects into a high-performance liquid chromatography (HPLC) system with a mass spectrometer (Davies, 2000). In addition to its use in animals, the microdialysis technique has been used to determine pesticide residue concentrations in plants (Zhou et al., 2009).

Another technique that shows some promise in this area is the use of solid-phase microextraction (SPME). This solventless technique was first introduced in the early 1990s but has found much of its use in the flavor and fragrance industry as well as for headspace analysis of volatile organic compounds. The way the method works is that a syringe is equipped with a retractable length of fiber that is coated with a bonded phase such as polydimethylsiloxane (PDMS). When the sample is exposed to the PDMS, chemicals within the sample will adsorb to the surface of the fiber. The fiber is then retracted and reinserted into the injection port of a gas chromatograph, where the chemicals are thermally desorbed onto the column (Ballesteros-Gómez, 2011). Recently a version of this technique, space-resolved SPME, was used to determine tissue-specific concentrations of atrazine as well as three pharmaceutical compounds in live rainbow trout (Zhang et al., 2010). The use of this, or other, passive sampling technique provides the advantage of avoiding loss of analytes during the cleanup and enrichment procedures prior to instrumental analysis.

Sample Extraction, Cleanup, and Enrichment

Once the samples have been collected, the analytes of interest (e.g., the pesticide and/or its metabolites) must be separated out from the matrix that was collected (e.g., the blood, urine, tissue, etc.). In most case this is done by bringing a suitable solvent into intimate contact with the sample, generally in a ratio of 5 to 25 volumes of solvent to 1 volume of sample or, more commonly, through the use of a solid-phase extraction (SPE).

The use of an electric blender is a common method of extraction of biological materials. The weighed sample is placed in a container, solvent is added, and the tissue is homogenized by motor-driven blades. Blending for 5 to 15 min followed by a repeat blending will extract most pesticide residues. A homogenate in an organic solvent can be filtered through anhydrous sodium sulfate to remove water that might interfere with the quantification phase of the analysis. The use of sonication is another method for extracting tissue samples, particularly when the binding of toxicants to subcellular fractions is of interest. Sonicator probes rupture cells rapidly, thus allowing the solvent to come into intimate contact with all cell components. In solid-phase extraction, liquid samples are filtered through a cartridge or filter disc made of material such as C-18 or Oasis HLB™ (hydrophilic-lipophilic balance). Analytes of interest are retained on the SPE filter and can be collected by eluting with various solvents or solvent mixtures. For example, the Oasis HLB™ can be used for the extraction of 29 currently used pesticides in human serum or plasma (Martínez Vidal and Garrido Frenich, 2006).

Using one of the extraction techniques described above will remove the analytes of interest from the bulk medium, but will also extract other matrix constituents (waxes, lipids, inorganic components, etc.). These interfering compounds must be removed prior to analysis, and there are various methods available to separate the desired components from the matrix interferences. A little over 100 years ago, a Russian botanist by the name of M.S. Tsweet published a paper describing a new method for separating out plant pigments by percolating a plant extract through a column of $CaCO_3$

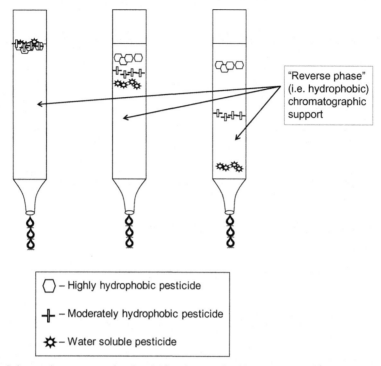

"Reverse phase" (i.e. hydrophobic) chromatographic support

◯ – Highly hydrophobic pesticide

╬ – Moderately hydrophobic pesticide

✷ – Water soluble pesticide

Figure 2.1 Column chromatography. Pesticides that are highly hydrophobic will have a high affinity for the stationary phase and will be the last to elute from the column, while water-soluble pesticides will elute first.

with petroleum ether. He called the process chromatography, or "color writing". The fundamental principles of chromatography are the same, regardless of whether it is preparative chromatography or analytical chromatography. There is a stationary phase and a mobile phase, and the separation of sample components is achieved through the differential interaction between the two phases. Compounds that have a higher affinity for the mobile phase will move through the system very quickly, whereas compounds that have a higher affinity for the stationary phase will move very slowly through the system (Figure 2.1). This process is the basis for the vast majority of analytical chemistry techniques in use today.

Thin-Layer Chromatography

Many toxicants and their metabolites can be separated from interfering substances with TLC. In this form of chromatography, the adsorbent is spread as a thin layer (250–2000 μm) on glass or resistant plastic backings. When the extract is placed near the bottom of the plate and the plate is placed in a tank containing a solvent system, the solvent migrates up the plate, and the toxicant and other constituents move with the solvent; differential rates of movement result in separation. The compounds can be scraped from the plate and eluted from the adsorbent with suitable solvents (Hodgson, 2010).

Column Adsorption: Hydrophobic and Affinity

A large number of adsorbents are available to the analyst. The adsorbent can be activated charcoal, aluminum oxide, Florisil, silica, silicic acid, or mixed adsorbents. The characteristics of the toxicant determine the choice of adsorbent. When choosing an adsorbent, select conditions that either bind the coextractives to it, allowing the compound of interest to elute, or vice versa. The efficiency of separation depends on the flow rate of solvent through the column (cartridge) and the capacity of the adsorbent to handle the extract placed on it. This amount depends on the type and quantity of adsorbent, the capacity factor and concentration of sample components, and the type and strength of the solvents used to elute the compound of interest (Hodgson, 2010).

Cartridge technologies are improving, however, to allow similar concentrations of sample to be added, which results in a less expensive and more rapid analysis. A number of SPE apparatuses have been introduced since the early 1980s. Most contain 0.5 to 2.0 g of the adsorbent in a plastic tube with fitted ends. The columns can be attached to standard syringes. Other companies have designed vacuum manifolds that hold the collecting device. The column is placed on the apparatus, a vacuum is applied, and the solvent is drawn through the column. Some advantages of these systems include preweighed amounts of adsorbent for uniformity, easy disposal of the coextractives remaining in the cartridge, no breakage, and decreased cost of the analysis because less solvent and adsorbent are used. Affinity chromatography is a potent tool for biologically active macromolecules that can be used for purifying small molecules, such as pesticides (Hennion, 2003; Stoks et al., 1999). It depends on the affinity of an enzyme for a substrate (or substrate analog) that has been incorporated into a column matrix or the affinity of a receptor for a ligand.

Size-Exclusion Chromatography

Also referred to as gel-permeation chromatography (GPC), this technique is primarily used during the analysis of biological samples. When tissue samples are extracted with a nonpolar solvent, significant amounts of lipids are also extracted. GPC columns are packed with a cross-linked polymer material that is very porous. Cross-linked dextrans such as Sephadex or agarose are commonly used materials. Small molecules can get into the pores and are thus retained for longer periods of time on the column, whereas large molecules (e.g., lipids) cannot and therefore pass through the column very quickly. The GPC material is available in varying pore sizes depending on the application for which it will be used. When this type of cleanup is used, the lipid fraction is retained, dried, and weighed. Thus, analytical results from the sample can be reported on a lipid weight basis where appropriate.

Enrichment

Finally the samples are concentrated to a very small volume of solvent, e.g., 250 µl, and appropriate standards are added for quality assurance and quality control purposes.

Analysis

Once the samples have been extracted, cleaned up, and concentrated, the next step is to analyze the extract using a variety of techniques, depending on the analytes of interest.

Gas Chromatography

GC is used most commonly for the separation and quantification of thermally stable pesticides such as organochlorines or pyrethroids. This system consists of an injector port, oven, detector, amplifier (electrometer), and supporting electronics. Gas chromatographs use a capillary column to effect separation of complex mixtures of organic molecules. The stationary phase is coated onto the inside of the capillary column. The mobile phase in this system is an inert gas (called the carrier gas), usually helium or nitrogen, that passes through the column. This technique is actually "gas–liquid chromatography," deriving from the fact that the polymer coating that acts as the stationary phase is technically a liquid.

Briefly, a sample is injected into a port that is at a temperature sufficient to vaporize the sample components (generally ≈300°C). Based on the solubility and volatility of these components with respect to the stationary phase, the components separate and are swept through the column by the carrier gas to a detector, which responds to the concentration of each component. The column is contained within an oven and the temperature within the oven can be programmed by the analyst. Similar to the way the solvent systems can be changed in column chromatography, the temperature program can be altered to maximize the analyte separation while minimizing the run time per sample. The electronic signal produced as the component passes through the detector is amplified by the electrometer, and the resulting signal is sent to a computer, or other electronic data-collecting device, for quantification. The time at which a specific compound exits the column for a given set of conditions within the instrument is called the retention time. Standard mixtures are run under the given conditions to determine the retention time for each analyte of interest. This is then compared with the retention times of peaks in the unknown samples.

Increased sensitivity and component resolution have resulted from advances in solid-state electronics and column and detector technologies. In the field of column technology, the capillary column has revolutionized pesticide detection in complex samples. This column generally is made of fused silica, 5 to 60 m in length, with a very narrow inner diameter (0.23–0.75 mm) to which a thin layer (e.g., 1.0 μm) of polymer is bonded. The polymer acts as the stationary phase. The carrier gas flows through the column at rates of 1 to 2 ml/min. Two types of capillary columns are used: the support-coated, open tubular (SCOT) column and the wall-coated, open tubular (WCOT) column. The SCOT column has a very fine layer of diatomaceous earth, coated with the stationary phase, that is deposited on the inside wall. The WCOT column is pre-treated and then coated with a thin film of stationary phase. Of the two columns, the SCOT is claimed to be more universally applicable because of large sample capacity, simplicity in connecting it to the chromatograph, and lower cost. However, for difficult separations or highly complex mixtures, the WCOT is more efficient and is used to a much greater extent (Hodgson, 2010).

High-Performance Liquid Chromatography

HPLC has become very popular in the field of analytical chemistry for the following reasons: it can be run at ambient temperatures; it is nondestructive to the compounds of interest, which can be collected intact; in many instances, derivatization is not nec-essary for response; and columns can be loaded with large quantities of material for detection of low levels. However, the most important advance was the development of an MS detector that could be coupled with HPLC.

The instrument consists of a solvent reservoir, gradient-forming device, high-pressure pumping device, injector, column, and detector. The principle of opera-tion is very similar to that of GC except that in this case the mobile phase is a liquid instead of a gas. The composition of the mobile phase and its flow rate effect separa-tions (recall that in GC, the oven temperature was controlled to effect separation). The columns being developed for HPLC are too numerous to discuss in detail. Most use finely divided packing (3–10 μm in diameter), some have bonded phases, and others are packed with alumina or silica. The columns normally are 15 to 25 cm in length, with small diameters (≈4.6 mm diameter). A high-pressure pump is required to force the sol-vent through this type of column.

Capillary Electrophoresis (CE)

A relatively new analytical technique, CE, is receiving considerable attention in the field of toxicology, and methods have been developed to analyze a diversity of com-pounds, including DNA adducts and pesticides. Commercial instruments are available that are composed of an autosampler, high-voltage power supply, two buffer reservoirs, the capillary (approximately 70 cm × 75 μm in diameter), and a detector. The versatility of the process lies in its ability to separate compounds of interest by a number of modes, including affinity, charge/mass ratios, chirality of the compounds, hydrophobicity, and size. The theory of operation is simple. Because the capillary is composed of silica, silanol groups are exposed on the internal surface, which can become ionized as the pH of the eluting buffer is increased. The ionization attracts cations to the silica surface, and when current is applied, these cations migrate toward the cathode, which causes a fluid migration through the capillary. This flow can be adjusted by changing the dielectric strength of the buffer, altering the pH, adjusting the voltage, or changing the viscosity.

Under these conditions both anions and cations are separated in a single separation, with cations eluting first. Neutral molecules (e.g., pesticides) can be separated by adding a detergent (e.g., sodium dodecyl sulfate) to the buffer, forming micelles into which neutral molecules will partition based on their hydrophobicity. Because the micelles are attracted to the anode, they move toward the cathode at a slower rate than does the remainder of the fluid in the capillary, thus allowing separation. This process is called micellar electrokinetic capillary chromatography. Many of these analyses can be carried out in 5 to 10 min with sensitivities in the low parts per billion range.

Mass Spectroscopy

Although there are many types of detectors available, especially for GC, the primary type of detector used for studies such as pesticide biotransformation is the mass spectrometer. The mass spectrometer is an outstanding instrument for the identification of a wide range of compounds. It is widely used as a highly sensitive detection method for GC and is increasingly used with HPLC and CE because technological hurdles have been solved, allowing these instruments to be interfaced with a mass spectrometer. Chromatographic techniques (e.g., GC, CE, HPLC) are used to separate individual components as previously described. A portion of the column effluent passes into the mass spectrometer, where it is bombarded by an electron beam. Electrons or negative groups are removed by this process, and the ions produced are accelerated. After acceleration they pass through a magnetic field, in which the ion species are separated by the different curvatures of their paths under gravity. The resulting pattern is characteristic of the molecule under study. Although there are many varieties of MS instruments, they share four common components: a system to introduce the sample, a method to produce ions, a method to separate and resolve the charged particles, and a detector. For example, a sample extract is injected into a gas chromatograph and the compounds are eluted into the mass spectrometer, the individual compounds are bombarded by an ionization source, and the resulting fragments are separated within the detector based on their mass-to-charge ratio (m/z). The resultant output, or the mass spectrum of a given compound, is unique, much like a fingerprint. Of the four components listed above, we have already discussed the first one. With the exception of the use of a direct injection probe to analyze the sample, MS is generally paired with an analytical instrument, such as a gas chromatograph or an HPLC apparatus to achieve separation of multiple analytes. So let us now turn our attention to the various methods of ionization. The most well established method, especially with regard to volatile samples, is electron–impact (EI) ionization. In this type of instrument, the mobile phase containing the sample is passed through a beam of highly energetic electrons ($\approx 70\,eV$), which are generated from a filament. This causes fragmentation of the molecule in a pattern that is indicative of the compound. In addition, some of the molecules may remain intact, producing a molecular ion, from which molecular weight is determined. The mass spectra produced by EI are fairly reproducible and there are large databases available with libraries of EI mass spectra.

With some compounds, the use of EI is too energetic to produce a molecular ion, and in these cases a "softer" method can be used: chemical ionization (CI). Chemical ionization can either be negative chemical ionization or positive chemical ionization (PCI), but in both cases, the ionization is achieved through the use of a reagent gas such as methane. For example, in PCI, the gas–phase reactions with methane form a strong acid, CH_5^+, which can then protonate the sample molecule, forming the molecular ion.

The previous two methods, EI and CI, are primarily used with GC instruments and thus are restricted to compounds that are thermally stable. However, compounds that decompose upon heating, or are polar and/or very large are more amenable to being run on an HPLC apparatus, which requires a different type of system to produce ions. Much of this is because the amount of solvent exiting the LC column is far more than exits the GC column. One common method is electrospray ionization (ESI). In this method, the solvent exits the LC column via a needle that has a high-voltage charge applied to it. This produces charged droplets that contain solvent molecules as well as the analyte molecules. As the droplets evaporate, charged molecules are ejected from the droplet. This technique is particularly useful for large molecules such as proteins. Other methods include atmospheric pressure chemical ionization (APCI) and thermospray ionization. Ionization with APCI is similar to ESI, but instead of an energized needle, the solvent enters a 500°C tube and is nebulized with an inert gas such as nitrogen. Ions are then formed in the plasma as they exit the tube. This is not a good choice for compounds that are thermally unstable. Finally, the use of a laser light can provide a soft ionization method. The most commonly known is that of matrix-assisted laser desorption (MALDI). This method is not coupled with an inlet such as GC or HPLC, but rather is used as a more direct method. The sample is embedded in a matrix that absorbs light at a λ near that of the laser, thus protecting the biomolecule from being broken apart. However, the energy ionizes the molecules in the sample.

Once the sample has been separated chromatographically and the analytes have been broken into ionized fragments, those fragments must be separated with a mass analyzer. The mass analyzer is that part of the instrument that sorts through all of the molecular ions and fragments that have been generated by the ionization source of the instrument. Generally speaking, there are two ways that the instrument can do this. The first is to continuously scan for every possible ion fragment all the time. This is sometimes called full-scan mode, and this method is useful when scanning matrices for a broad range of possible unknown compounds. Its disadvantage is that the sensitivity of the method will be reduced. In other cases, however, one might know exactly what one is looking for, such as pyrethroid and pyrethrin metabolites (Martínez Vidal and Garrido Frenich, 2006). In this case one would use selected-ion monitoring (SIM). Using SIM, the analyst programs the instrument such that, at a given retention time, the mass analyzer will scan for only a few specific ions. In this way the amount of dwell time for each ion can be greatly increased and therefore the sensitivity of the method is increased as well. Again there are several different types of mass analyzers to consider.

The real workhorse in this area is the quadrupole, as it is compact, robust, and relatively inexpensive. As the name implies, the basic configuration consists of four rods that are arranged in parallel around a central channel. A fluctuating electric field is formed in this channel by applying DC and RF voltages across opposing pairs of rods. By modulating the currents, only ions with a particular m/z ratio can traverse the channel to reach the detector at the opposite end. More recently, these quadrupole instruments have been set up in tandem to derive much more structural information from the samples. These instruments, generally referred to as a triple-quad or MS/MS instruments, use a technique called collision-induced dissociation. In this case a particular ion passes through the first quadrupole, then it enters a mixing chamber where more ions are formed through a chemical ionization technique, and then these resultant ions enter a second quadrupole mass analyzer. These instruments are particularly useful in biological applications and are very commonly paired with HPLC.

A newer design that is rapidly becoming more prevalent is the ion trap (Ballesteros-Gómez, 2011). The principle of ion separation for the ion-trap instrument is similar to that for the quadrupole, but rather than taking place along a linear channel, it occurs in an inner-tube-shaped ring with caps on the end. This circular magnet "traps" the ions and they are then swept out by RF voltage. The advantage of the ion trap is that it can hold the ions in place and perform the MS/MS analysis in situ, without having a second mass analyzer. Therefore, the analyst is not constrained by MS/MS, but rather could repeat the process to achieve MS/MS/MS/.../MS^X. A new mass analyzer that is similar in principle to the ion trap is the Orbitrap. This mass analyzer traps the ions around a central spindle electrode and the m/z values are measured from the frequency of harmonic ion oscillations, along the axis of the electric field (Hu et al., 2005).

The final method to address here is the time-of-flight (TOF) analyzer. Essentially the TOF works on the principle that, if ions are accelerated through a fixed distance by a high voltage, larger ions will move more slowly than smaller ions. Thus, the ions are resolved based on the amount of time it takes them to traverse the tube. This analyzer is often paired with the MALDI technique for the determination of large molecules.

UPTAKE, DISTRIBUTION, AND TOXICOKINETICS

One of the central aims of toxicology in general, and pesticide disposition/biotransformation in particular, is the determination of the absorption, distribution, metabolism, and elimination of the pesticide. Increasingly this is assisted by the use of physiologically based pharmacokinetic (PBPK) models. These models divide the organism into discrete compartments with mathematical expressions to describe the affinity of the pesticide for the compartment as well as the partitioning between compartments. This is discussed in further detail in Chapters 3 and 6 of this book. However, to build a physiologically based model, one has to have physiological parameters based on experiments in vivo. Dosing studies carried out in rodents can use a variety of dosing techniques, including oral dosing, inhalation, percutaneous absorption, intravenous (iv) injection, or iv infusion.

When using an oral dosing design there are a number of considerations to take into account. First, because of the various biological barriers to absorption in place, a larger dose can generally be administered compared with an iv administration. In general, most laboratory animals may be successfully dosed by gavage, with rabbits being the exception. Other important considerations include the composition, pH, and volume of the solution containing the pesticide. Ideally, the vehicle should be at neutral pH, and different dosing levels should still use the same total volume of vehicle to avoid artifacts in the data. In general, a vehicle volume of less than 10 ml/kg will not cause interference with uptake kinetics.

Determining uptake via inhalation is more problematic than oral exposure, the primary challenge being the differentiation of the true inhalation exposure from that which is swallowed (e.g., through grooming) or absorbed dermally. This can be remedied through the use of nose-only exposures (Langenberg et al., 1998; Pauluhn and Mohr, 2006). For example, Yoshida et al. (1991) observed a difference in the calculated LC_{50} of Chloropicrin in rats exposed via nose-only exposure versus whole-body inhalation exposure.

The dermal route of exposure can be an important consideration for pesticides, given the nature of work performed by agricultural and commercial pesticide applicators. There are many considerations in the choice of animal model. On one hand, most disposition and metabolism studies are carried out in mice and rats, so for comparability between routes of administration one would favor these models. However, the structure and functions of the skin in humans are very different from those of rodents, and thus the animal models with the closest physiology to human skin would be some primates as well as the pig. One technique to determine uptake of pesticides in intact skin is through the use of the isolated perfused porcine skin flap (Riviere et al., 1986; Chang et al., 1994). This device uses a surgically prepared vascularized flap of porcine skin, which can be removed and placed into a chamber. The intact vasculature can be attached to a pump and the perfusate monitored for the presence of the compound being tested.

In determining PBPK parameters such as the apparent volume of distribution, the data from an iv injection are one of the most important factors. As with oral dosing, the volume of the vehicle is an important consideration. In general 1 to 2 ml/kg, depending on whether it is an aqueous or a solvent vehicle, is appropriate. Although the tail and hindpaw veins are adequate, the ideal administration is through the use of a cannula inserted into the femoral or external jugular vein. Finally, several important factors must be considered. First, in using a cannulated animal, care must be taken that the affinity of the pesticide for the cannula material is low, so that adsorption to the cannula material does not affect the results. Second, when these data go into a PBPK model, the assumption is that the time of injection is t_0. However, too rapid an administration can result in acute toxicity to the animal, and thus a slower injection may be required. Provided that the total time of injection does not exceed 5% of the half-life of the most rapid phase of the plasma concentration–time curve, it can be regarded as an instantaneous dose (Hayes, 2008). Finally, care must be taken that none of the administered dose ends up in perivascular compartments. In addition to bolus dosing, one can also carry out studies using iv infusion. The development of implantable osmotic mini-pumps has allowed the use of longer term pharmacokinetic studies (Sai et al., 2009).

In many ways, the advances in PBPK models were made possible by an electrical engineer working at Texas Instruments in the 1950s. In 1959 Jack Kilby filed a patent application for the integrated circuit, and without the integrated circuit there would not be ubiquitous and powerful computers, and without increasingly powerful computers, the process of PBPK modeling would be prohibitively complex. There are free, Web-based tools that can be used to build pharmacokinetic models, such as GNU MCSim and PBPK.org. In addition there are many models that are available commercially, including AcslIX, Cloe PK, and PK Sim. Again, the reader is referred to Chapter 6 of this text for a more complete discussion of the process.

CELL CULTURE, SUBCELLULAR FRACTIONS, AND RECOMBINANT ENZYMES

Cell culturing techniques have facilitated great strides in toxicology research. The two main limitations of these techniques are the changes that occur during isolation and propagation in culture, and the difficulty of duplicating kinetic aspects of toxicant exposure that occurs in the intact animal (Hodgson, 2010).

One method of avoiding the issue of the loss of complexity with a cellular monoculture is the use of tissue slices. These are $\approx$200-μm slices that are prepared using vibrating, precision-slicing instruments; however, their viability is limited to short-term studies.

With the exception of cells circulating (such as blood) or easily obtained by lavage (such as peritoneal or alveolar macrophages), isolation of cells requires a method to remove the cells from the solid tissue. Because (in general terms) the processes that hold cells together are Ca^{2+} dependent, the separation of cells generally involves removal of Ca^{2+} with a chelating agent such as ethylenediaminetetraacetic acid and the use of a proteolytic enzyme such as trypsin. Many methods can be used to separate the various cell types of interest, including centrifugation within a density gradient, which will separate the cells based on size; use of media that will favor the growth of a given cell type (e.g., hepatocytes are the only cell type that will grow in arginine-free media); or use of separation techniques such as the fluorescence-activated cell sorter. In general, a beam of a single-wavelength laser light is directed onto a stream of fluid containing the suspension. Forward-scatter, side-scatter, and fluorescence detectors receive input from the focused light. The suspended particle in solution scatters the light, and fluorescent tags, found in the particle or attached to the particle, can be detected by the instrument. Essentially, the instrument can determine the cell volume, as well as other features of the particle such as the membrane roughness, the shape of the nucleus, or the amount of cytoplasmic granules.

In general, suspension cultures tend to be short-term cultures and thus the medium requirements are relatively simple: salts to maintain osmolarity, a pH buffer, and an energy source, e.g., NaCl, bicarbonate, and glucose. In addition, a gas phase of CO_2 and O_2 is necessary. For longer-term research needs, monolayer cultures are used. In general terms, the tissue is cultured onto a polystyrene culture dish, as it has a charged surface to which the cells can initially attach. They will then generate an underlying matrix and will ultimately spread to form a monolayer. Growth and maintenance for more than a few hours requires additional nutrients including essential amino acids, vitamins, additional salts, and trace minerals. Initially there is "log-phase" or exponential growth, until the dish is filled (confluence), and then contact inhibition restricts further growth. This is referred to as a primary culture. Subculturing involves taking all or a portion of this and moving it to other vessels. Eventually cells stop dividing, leading to replicative senescence. Occasionally, however, some cells will continue to replicate and overtake the plate, which can lead to the development of continuous cell lines. These so-called "immortalized cells" have a loss of contact inhibition, which is evident in the small piles of cells (or foci) that overlie the monolayer. This anchorage-independent growth is operationally defined as the ability to grow in soft agar. Stem cells are increasingly being used in toxicology studies as new and better methods of using these cells are developed. One of the most versatile uses of stem cells is in the creation of so-called knockout mouse models, in which an intact animal can be reared with a given gene of interest removed.

Development of cell-permeative fluorescent tags has provided a means to observe alterations in cell function, as well as morphological changes. Inverted fluorescence microscopes are used to capture images using cells that have been treated with fluorescent tags. Tags are available to monitor such endpoints as oxidant status; sulfhydryl content; intracellular Ca^{2+}, H^+, Na^+, and K^+; mitochondrial function; and membrane potential. Digital electronic imaging and computerized data analysis can enhance the sensitivity of this technique and provide information on temporal relationships between multiple responses within a given cell. The disadvantage is poor resolution. However, this can be overcome through the use of laser-scanning confocal light microscopy, which can optically limit the image to thin slices within the depth of the monolayer.

The use of various molecular biological methods has greatly increased the pace of research in the biomedical sciences in general, and toxicology in particular. One common method is the use of recombinant DNA technology and molecular cloning. This is a method whereby strands/sequences of DNA that one would not normally encounter can be created and multiplied using vectors. A vector is simply a DNA molecule that originates from a plasmid or a virus that has the capacity to self-replicate when placed into a host cell. In this manner, hundreds of exact copies of the DNA strand of interest are produced within the cell. These cloned genes can then be used for a variety of purposes. The gene could be sequenced to determine the amino acid sequence that would be generated from the DNA or it could be put into a mammalian expression vector and expressed in mammalian cells.

PROTEOMICS

The proteome can be defined as that portion of the genome that is expressed as proteins in a cell or organism over time. Proteomics is a global analysis to study the structure and function of the proteome, involving its separation and identification. Finally, toxicoproteomics is the study of proteomics applied to toxicology in such a way as to identify critical proteins and pathways affected by toxicant exposure. There are seven attributes of proteins needed for comprehensive protein expression analysis: identity, quantity, post-translation modification, structure, protein-protein interactions, cellular-spatial relationships, and function. Given that proteins such as enzymes and cellular receptors are the preferred targets for virtually all pesticides, the field of proteomics promises to contribute significantly to the study of pesticide toxicology. Large-scale, high-throughput omics technologies are increasingly being used to elucidate highly complex networks of proteins and protein interactions within a biological system (Figure 2.2). The methods employed in these complex studies include two-dimensional gel electrophoresis, chromatographic separations, stable isotope labeling, mass spectroscopy, and protein bioinformatics.

In its simplest form, two-dimensional gel electrophoresis (2D-GE) involves running out proteins on a gel, rotating the gel 90°, and then further separating the proteins by a different property. For instance, proteins may be separated by mass and then

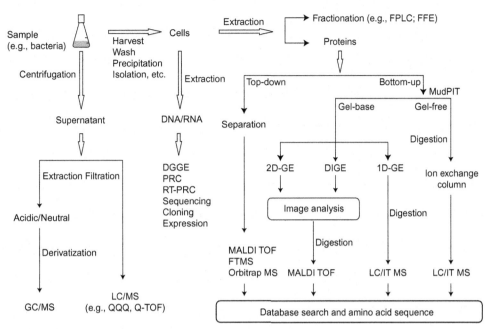

Figure 2.2 Schematic of "multiomics" approaches applied in metabolism studies of pesticides, in which MS-based proteomics and metabolomics workflow is emphasized. Metabolites in the supernatant of bacterial samples can be analyzed by gas chromatography/mass spectrometry (GC/MS) and liquid chromatography/mass spectrometry (LC/MS). DNA and RNA extracted from cells can be analyzed by denatured gradient gel electrophoresis (DGGE), polymerase chain reaction (PCR), real-time PCR (RT-PCR), sequencing, cloning, and other molecular techniques. Proteins extracted from bacterial cells can be analyzed via top-down or bottom-up approaches. In the bottom-up approach, a mixture of proteins can be separated by sodium dodecyl sulfide–polyacrylamide gel electrophoresis and then analyzed by MS. Alternatively, the protein mixture can be directly digested into a collection of peptides that are then separated and determined by multidimensional chromatography online coupled to tandem mass spectrometric analyses [i.e., multidimensional protein identification technology (MudPIT)]. In the top-down approach, intact proteins are fractionated into less complex protein mixtures for MS analysis. Bioinformatics including database search and amino acid sequence alignment is then conducted to identify and characterize the proteins and peptides. DIGE, difference gel electrophoresis; FFE, free-flow electrophoresis; FPLC, fast protein liquid chromatography; FTMS, Fourier transform mass spectrometry; MALDI TOF, matrix-assisted laser desorption ionization time-of-flight; QQQ, triple quadrupole; Q-TOF, quadrupole time-of-flight. *(Adapted from Hayes and Krieger, 2010.)*

separated by isoelectric point. Recent improvements in 2D-GE techniques include the development of immobilized pH gradients, very sensitive staining methods, and more powerful image analysis systems. A related technique, differential in–gel electrophoresis, has improved the reproducibility of 2D–GE by minimizing variations between gels. These methods are commonly combined with MALDI-TOF MS (as discussed under Analysis).

The shotgun approach to proteomics is an alternative to the electrophoretic methods discussed above. In this method, complex mixtures of proteins are analyzed using some of the many methods described earlier in this chapter, including LC/MS and LC/MS/MS. Protein quantitation can be a drawback to this methodology. In addition, 1D-GE can be combined with LC separation whereby a mixture of proteins is separated on a 1D gel and individual bands are digested with trypsin. The tryptic peptides are, in turn, analyzed on LC/MS and MS/MS. There are numerous commercial software packages available to analyze the resultant data (Hayes and Krieger, 2010).

Quantification of proteins is an active area of proteomics research and can be achieved through the use of stable isotope probes. Labeling techniques, such as the isotope-coded affinity tag or the cleavable isotope-coded affinity tag, allow proteins to be separated on avidin/biotin columns but label cysteine residues exclusively. Isobaric tags for relative and absolute quantitation allow for amine-specific isobaric tagging and thus are suitable for simultaneous analysis of proteins regardless of the presence of cysteine residues. One novel method can label arginine and lysine in vivo, stable isotope labeling by essential amino acid culture.

All of these techniques rely heavily on bioinformatics. Tools for bioinformatics include software packages to analyze the extremely large data sets that are derived from 2D-GE and MS analysis. Software packages are available to analyze 2D-GE images for the detection and semiquantification of protein spots on 2D gels, the localization of protein spots within a gel, the matching of corresponding spots between gels, and the differential comparison of protein expression. For a more complete list of the various models, databases, and software packages available, the reader is directed to Chapter 21 in Hayes and Krieger (2010).

METABOLOMICS

Metabolomics is the systematic study of a metabolome, the entirety of metabolites, or a set of metabolites, forming an extensive network of metabolic reactions in which one metabolite from a specific pathway will affect one or more biochemical reactions, or a comprehensive and quantitative analysis of all metabolites. Arguably, metabolomics is the least well characterized of all the omics in systems biology, as well as the most complex and dynamic. Consequently, it is perhaps less well defined. Indeed the term "metabolomics" was first proposed only in 1998, although various forms of research in metabolic profiling have been undertaken for many years. Given the wide variation in the possible metabolites of a given chemical such as a pesticide, there is no one unified method to determine the metabolome. In general terms, the three primary methods used in metabolomics studies include MS, nuclear magnetic resonance (NMR), and deconvolution and structure identification. As with proteomics, there is also significant reliance on bioinformatics and data mining.

Mass spectroscopy has previously been discussed in this chapter but another very powerful tool for determining structure information on molecules is NMR. Although not as sensitive as many of the MS technologies, NMR offers the advantage of rapid data acquisition and excellent reproducibility. However, improvements in instrument design (higher field strength magnets) as well as improvements in spectral interpretation software have made NMR one of the cornerstones of metabolomics research. Perhaps the most widely used method is ^{1}H NMR; however, other heteroatoms may be used as well, including ^{13}C, ^{31}P, and ^{15}N. The last can be used for multidimensional NMR analysis. One of the most noticeable applications of NMR is metabolomics with solid samples such as organs and cells. Magic angle spinning-NMR techniques are becoming common for solid-state sample analyses (Hayes and Krieger, 2010). Finally, with both MS and NMR techniques, there are often overlapping peaks within the spectra; deconvolution involves methods for teasing out the spectra of individual compounds (metabolites). There are several software tools and spectral libraries available to perform these analyses. For a more detailed treatment of this topic, the reader is referred to Hayes and Krieger (2010, Chapter 22).

SUMMARY

Regardless of the type of study, whether a simple disposition study or a large-scale metabolic profiling study, the essential elements are the same: defining the research goal(s), identifying the appropriate techniques and methods, developing a sampling scheme to obtain representative samples, isolating the compound(s) of interest, removing potentially interfering components, and quantifying and evaluating the data in relation to the original research goals. These essential elements are critical to ensuring that the data generated address the original hypothesis. This is especially true as more sophisticated techniques, such as PBPK models or proteomics, generate more and more massive data sets, as errors early on in the experimental design will propagate through the study. Meaningful data can be generated only if the proper method of analysis is employed prudently within a robust experimental design.

REFERENCES

Ballesteros-Gómez, A., & Rubio, S. (2011). Recent advances in environmental analysis. *Anal. Chem.,* *83*(12), 4579–4613.

Chang, S. K., Brownie, C., & Riviere, J. E. (1994). Percutaneous absorption of topical parathion through porcine skin: In vitro studies on the effect of environmental perturbations. *J. Vet. Pharmacol. Ther.,* *17,* 434–439.

Davies, M. (2000). Analytical considerations for microdialysis sampling. *Adv. Drug Delivery Rev.,* *45,* 169–188.

Hayes, A. W. (Ed.). (2008). *Principles and methods of toxicology* (5th ed.). Boca Raton, FL: CRC Press.

Hayes, W. J., & Krieger, R. I. (Eds.). (2010). *Hayes' handbook of pesticide toxicology.* Amsterdam: Elsevier/Academic Press.

Hennion, M. (2003). Immuno-based sample preparation for trace analysis. *J. Chromatogr. A, 1000,* 29–52.

Hodgson, E. (Ed.). (2010). *A textbook of modern toxicology*. Hoboken, NJ: John Wiley & Sons.

Hu, Q., Noll, R. J., Li, H., Makarov, A., Hardman, M., & Cooks, R. G. (2005). The Orbitrap: A new mass spectrometer. *J. Mass Spectrom., 40*, 430–443.

Kurien, B. T., Everds, N. E., & Scofield, R. H. (2004). Experimental animal urine collection: A review. *Lab. Anim., 38*, 333–361.

Langenberg, J. P., Spruit, H. E. T., van der Wiel, H. J., Trap, H. C., Helmich, R. B., Bergers, W. W. A., et al. (1998). Inhalation toxicokinetics of soman stereoisomers in the atropinized guinea pig with nose only exposure to soman vapour. *Toxicol. Appl. Pharmacol., 151*, 79–87.

Martínez Vidal, J. L., & Garrido Frenich, A. (Eds.). (2006). Pesticide protocols. In J. M. Walker (Ed.). *Methods in biotechnology* (Vol. 19). Totowa, NJ: Humana Press.

Mattila, S., Reponen, P., Abass, K., & Pelkonen, O. (2010). Overview of the metabolism and interactions of pesticides in hepatic in vitro systems. *Int. J. Environ. Anal. Chem., 90*, 429–437.

Pauluhn, J., & Mohr, U. (2006). Mosquito coil smoke inhalation toxicity. Part II. Subchronic nose-only inhalation study in rats. *J. Appl. Toxicol., 26*, 279–292.

Plock, N., & Kloft, C. (2005). Microdialysis—theoretical background and recent implementation in applied life-sciences. *Eur. J. Pharm. Sci., 25*, 1–24.

Riviere, J. E., Bowman, K. F., Monteiro-Riviere, N. A., Dix, L. P., & Carver, M. P. (1986). The isolated perfused porcine skin flap (IPPSF). *Fundam. Appl. Toxicol., 7*, 444–453.

Sai, Y., Chen, J., Wu, Q., Hiu, L., Zhao, J., & Dong, Z. (2009). Phosphorylated-ERK 1/2 and neuronal degeneration induced by rotenone in the hippocampus neurons. *Environ. Toxicol. Pharmacol., 27*, 366–372.

Stoks, P., Houben, A., Gronert, C., Meulenberg, E., & Noij, T. (1999). Immuno affinity extraction of pesticides from surface water. *Anal. Chim. Acta, 399*, 69–74.

Yoshida, M., Murao, N., Tsuda, S., & Shirasu, Y. (1991). Effects of mode of exposure on acute inhalation toxicity of chloropicrin vapor in rats. *J. Pestic. Sci., 16*, 63–69.

Zhang, X., Oakes, K. D., Cui, S., Bragg, L., Servos, M. R., & Pawliszyn, J. (2010). Tissue-specific in vivo bioconcentration of pharmaceuticals in rainbow trout (*Oncorhynchus mykiss*) using space-resolved solid-phase microextraction. *Environ. Sci. Technol., 44*, 3417–3422.

Zhou, S. N., Oakes, K. D., Servos, M. R., & Pawliszyn, J. (2009). Use of simultaneous dual-probe microdialysis for the determination of pesticide residues in a jade plant (*Crassula ovata*). *Analyst, 134*, 748–754.

Absorption

Ronald E. Baynes, Jim E. Riviere
North Carolina State University, Raleigh, NC, USA

Outline

INTRODUCTION

For a pesticide to elicit toxicity, it must be transferred from the external site of exposure to the target site (e.g., organ, nucleic acid, receptor) and achieve a sufficiently high concentration in the target organ (Figure 3.1). Absorption is the translocation of the pesticide from an external source of exposure to the bloodstream. Once in the blood, the chemical is distributed through the body and delivered to tissues, where it may leave the blood and enter the cells of the tissue or it may remain in the blood and simply pass through the tissue. In certain tissues such as the liver, the chemical may be effectively removed from the body by metabolism. Other tissues, such as kidney and lung, serve to eliminate xenobiotics from the body by excretion. Absorption, distribution, metabolism, and excretion, which are collectively termed disposition, are all factors that affect the concentration of a chemical in target tissues. Pharmacokinetics refers to the mathematical description of the time course of chemical disposition in the body. Metabolism and

Pesticide Biotransformation and Disposition
DOI: 10.1016/B978-0-12-385481-0.00003-4

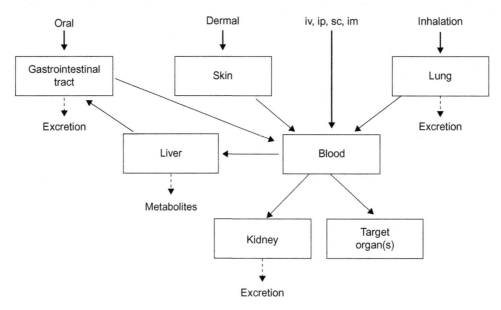

Figure 3.1 Representation of the absorption, distribution, metabolism, and excretion of toxicants.

excretion are discussed in detail in other chapters of this work. This chapter focuses on pesticide absorption with an expanded focus on dermal absorption, as workers involved in pesticide manufacturing, formulation, or application and harvesting of treated crops are more likely to be exposed to these chemicals via the skin.

FACTORS THAT INFLUENCE THE TRANSFER AND AVAILABILITY OF CHEMICALS IN THE BODY

For the routes of pesticide exposure relevant to humans, the pesticide must cross one or more cell membranes to reach the bloodstream, and then one or more additional cell membranes to leave the blood and enter tissues. The following discussion concerns the factors that influence the transfer of chemicals across biological membranes.

Properties of Cell Membranes

Cell membranes (i.e., plasma membranes) consist of phospholipids and proteins (Figure 3.2). The fluid and dynamic phospholipid bilayer, with polar head groups on the intracellular and extracellular surfaces and fatty acid chains filling the inner space, acts as a permeability barrier to water-soluble molecules. Proteins interspersed throughout the phospholipid bilayer mediate the transport of small, water-soluble molecules into and out of the cell by forming pores or by acting as carriers. Molecules

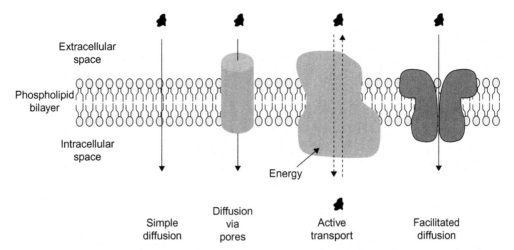

Figure 3.2 Schematic of the plasma membrane and mechanisms of transport across the membrane.

cross membranes by passive transport, which requires the expenditure of no energy, or by specialized transport systems. The ability of a chemical to cross various membrane barriers is determined by its physicochemical properties, which include lipophilicity, molecular size, and ionization.

Transport Mechanisms

Passive Transport

Passive transport occurs by simple diffusion or via pores in the plasma membrane (Figure 3.2). Most lipophilic molecules cross membranes by simple diffusion in accord with Fick's first law of diffusion (see equation below), which states that the flux or rate at which a molecule diffuses across the plasma membrane is proportional to the concentration gradient, the membrane surface area, and the permeability coefficient of the molecule. The permeability coefficient is the product of the partition coefficient and the diffusion coefficient.

$$\text{flux} = \frac{\text{diffusion coefficient} \times \text{surface area} \times \text{partition coefficient} \times \text{conc. gradient}}{\text{skin or membrane thickness}}$$

Theoretically, the determinants of flux or diffusion rates across the skin or gastrointestinal (GI) tract may be altered clinically or experimentally through manipulation of pesticide formulations. If lipid solubility increases, the penetrant may remain in the stratum corneum of the skin and form a reservoir. Some compounds can also form a reservoir in the dermis. These scenarios can prolong absorption half-life across the skin, which can also prolong the body burden of the penetrant. Ingestion of very lipid soluble pesticides, which are not miscible in the aqueous intestinal fluid, can be presented as emulsions and brought into solution through the action of detergent-like bile acids. The products of this mixing are large-surface-area micelles (hydrophobic interior) that deliver the lipids to the brush border of the intestine for diffusion across the membrane.

Water readily traverses the plasma membrane through pores and may carry with it small hydrophilic solutes. The pores in most cells are approximately 4 Å in diameter. In the kidney glomeruli, however, the pores may be as large as 70–80 Å in diameter, which permits more efficient renal elimination of potentially toxic compounds. Weak organic acids and bases may cross plasma membranes by simple diffusion when they are non-ionized. Ionized weak organic acids and bases, however, slowly permeate the plasma membrane through pores.

According to the Brønsted-Lowry theory, an acid is a proton donor and a base is a proton acceptor. The ratio of non-ionized to ionized molecules of a weak organic acid or base depends on the dissociation constant (K_a) and the pH of the medium (Table 3.1). The dissociation constant is usually expressed in terms of its negative logarithm, and the relationship between pK_a and pH is derived from the Henderson-Hasselbalch equation as shown in Table 3.1. The pK_a is the pH at which 50% of the acid or base is ionized. The concept of pK_a is particularly important for oral absorption (see Absorption from the Gastrointestinal Tract) and often overlooked when assessing dermal absorption (see Percutaneous Absorption).

The penetration of acidic and basic pesticides through skin can be influenced by the skin surface pH, which is weakly acidic (pH 4.2–5.6). Paraquat and diquat are hydrophilic pesticides that exist as fixed charged cations and remain dissociated at all pH values. Very little paraquat or diquat is, therefore, expected to be absorbed by skin, although percutaneous absorption of paraquat has resulted in systemic effects and deaths in humans (Smith, 1988). Dermal absorption studies in human volunteers demonstrated 0.29, 0.23, and 0.29% absorption in the leg, forearm, and forearm, respectively (Wester

Table 3.1 Acids and Bases According to the Brønsted-Lowry Theory

	Acid	Base
Representation	$AH \leftrightarrow A^- + H^+$	$B + H^+ \leftrightarrow BH^+$
Definition	Proton donor (AH)	Proton acceptor (B)
Dissociation constant (K_a)	$K_a = \dfrac{[A^-][H^+]}{[AH]}$	$K_a = \dfrac{[B][H^+]}{[BH^+]}$
$pK_a = \log \dfrac{1}{K_a} = -\log K_a$	$pK_a = pH + \log \dfrac{[AH]}{[A^-]}$	$pK_a = pH + \log \dfrac{[BH^+]}{[B]}$

et al., 1984). Other studies have determined that the in vitro permeability constants for paraquat in various animal species (rat, hairless rat, nude rat, mouse, hairless mouse, rabbit, guinea pig) are 40–1600 times greater than for humans (Walker et al., 1983). One radiolabeled in vivo rat study reported a dermal bioavailability of 3.8% (Chui et al., 1988), which supports the claim that rodent studies can overestimate human absorption. Like paraquat, very little diquat is absorbed (0.3%) in the human forearm in vivo (Maibach and Feldmann, 1974). Diquat absorption increased to 1.4% with occlusion and to 3.8% with damaged skin. Data from these in vivo *and* in vitro studies suggest that paraquat- or diquat-induced dermatotoxicity is a highly probable mechanism, a priori, for dermal absorption of these hydrophilic and charged pesticides.

Specialized Transport

Active transport systems are characterized by (1) movement of solutes against a concentration or electrochemical gradient, (2) saturation at high solute concentration, (3) specificity for structural and/or chemical features of the solute, (4) competitive inhibition by molecules transported by the same transporter, and (5) inhibition of transport by compounds and/or processes that interfere with cellular metabolism. Facilitated diffusion is similar to active transport, except that the solute moves only in the direction of a concentration or electrochemical gradient and the expenditure of energy is not required (Figure 3.2). Additional types of specialized transport are exocytosis and endocytosis, processes by which cells secrete and ingest large molecules, respectively. There are two types of endocytosis: pinocytosis (cell drinking), which is the ingestion of fluids and solutes, and phagocytosis (cell eating), which is the ingestion of large particles. Phagocytosis is especially important in the removal of particulate matter in the respiratory tract. Recent studies have also suggested an even finer gradation in specific transport processes (e.g., caveolae) that facilitate entry of different-sized material into the cell.

Many of the available commercial pesticides are transported across the skin and GI tract by passive diffusion. However, there is some evidence that membrane transport proteins play a significant role in the absorption mechanism in the GI tract and account for pesticide influx and/or efflux of several pesticides. The hydrophilic herbicide paraquat is thought to be absorbed by a mechanism that consists of facilitated, saturable, and diffusional components (Heylings, 1991; Nagao et al., 1993). The P-glycoprotein (P-gp/MDR1) is a transmembrane transporter in humans and animals that is encoded by the *ABCB1/MDR1* gene. This transporter is in various human tissues such as the apical surface of intestinal epithelial cells. The interactions between P-glycoproteins and the avermectin class pesticides and other classes of insecticides such as methylparathion, endosulfan, cypermethrin, and fenvalerate have been well documented (Sreeramulu et al., 2007; Zhou, 2008). A similar or related mechanism has been reported for the influx and efflux of neonicotinoids (Brunet et al., 2008). These interactions are important as they dictate the rate and extent of pesticide absorption across the intestine, especially when there is coexposure to other drugs or pesticides that may compete with the pesticide of interest and consequently increase pesticide uptake across the GI tract (Alvinerie et al., 2008).

Protein (Macromolecular) Binding

Blood consists of red blood cells, white blood cells, and platelets suspended in plasma. Plasma, which comprises approximately 55% of the blood volume in humans, also contains a number of proteins, ions, and inorganic molecules. Many xenobiotics in blood are reversibly bound to plasma proteins, including albumin, α_1-acid glycoprotein, lipoproteins, and globulins. Reversible binding to plasma proteins enhances the solubility of lipophilic compounds in blood and influences the rate of distribution to tissues. Proteins are amphoteric in nature and therefore possess cationic and anionic regions. Many acidic chemicals bind to albumin, whereas basic chemicals tend to bind to α_1-acid glycoprotein and lipoproteins. The high molecular weight of proteins prevents them, and any toxicants they bind, from crossing cell membranes. Only the free (or unbound) chemical is available to cross plasma membranes (Figure 3.3). The interaction of chemicals and plasma proteins, however, is rapid and reversible. Equilibrium is quickly established between the bound and the unbound forms of the chemical. As an unbound chemical crosses a plasma membrane in a microenvironment, a bound chemical dissociates to re-establish equilibrium with the unbound fraction.

Gomez–Catalan et al. (1991) investigated the distribution of various organochlorines in rat and human blood. In rat blood, 87% of hexachlorobenzene was associated with red blood cells, approximately 84% of DDE 1,1–dichloro–2,2–bis(chlorophenyl)ethylene was bound to plasma proteins, lindane was nearly equally distributed between red blood cells

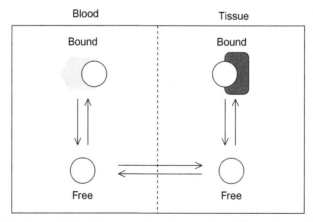

Figure 3.3 Equilibrium is established between free (unbound) and bound xenobiotics in blood and between free xenobiotics in blood and tissues. Only the free xenobiotic crosses the plasma membrane, which is represented by the dashed line separating the blood and tissue compartments.

and plasma, and 97% of pentachlorophenol was associated with plasma. In plasma, lindane (64%) and DDE (92%) were mainly associated with lipoproteins, pentachlorophenol was mainly associated with "other" plasma proteins (81%), and hexachlorobenzene was nearly equally distributed. A very different pattern of distribution was observed in human blood. Hexachlorobenzene and lindane in plasma were nearly equally distributed between lipoproteins and "other" plasma proteins, whereas 60% of DDE was associated with "other" proteins. Other investigators have shown that dieldrin is over 99% bound to human serum proteins (Garrettson and Curley, 1969), and diflubenzuron is 40–50% bound to plasma proteins in chickens (Opdycke and Menzer, 1984). The organophosphate diazinon is 89% bound to proteins in rat plasma (Wu et al., 1996).

ABSORPTION

Human exposure to pesticides is typically by oral, dermal, and inhalation routes. Occupational exposure to pesticides is more likely to occur by dermal contact, and this is the focus of the section Percutaneous Absorption. Percutaneous absorption is reported as the possible route of entry in 65–85% of all cases of occupational exposure to pesticides (Galli and Marinovich, 1987). Spray or dusting of pesticides can result in disposition of 20–1700 times the amount deposited in the respiratory tract (Feldmann and Maibach, 1974). Epidemics of pesticide poisoning following cutaneous exposure have been reported for nonoccupational uses (Ferrer and Cabral, 1993). These cases have often involved accidental contamination of infant clothing or exposure to talcum powder with pesticides (Martin-Bouyer et al., 1983). These anecdotal case reports, coupled with dermal exposure estimated from various direct and indirect dosimetric experiments, are often the only available human data with which to perform dermal absorption assessment. Despite such limited data, it is possible to estimate dermal absorption by extrapolation from dermal exposure data. Algebraic equations that take into account exposure time and the chemical nature of the compound (lipophilicity and molecular weight) have been presented for estimating dermal absorption (Cleek and Bunge, 1993; Potts and Guy, 1992).

Exposure of small children is more likely to be by oral and/or dermal routes. Absorption from the gastrointestinal and respiratory tracts is discussed under Absorption from the Gastrointestinal Tract and Absorption from the Respiratory Tract, respectively. Other routes of exposure that are used primarily in the laboratory (subcutaneous, intravenous, intraperitoneal, and intramuscular) are discussed only briefly under Absorption after Exposure by Other Routes.

A chemical is considered to be absorbed when it reaches the bloodstream. For routes other than intravenous administration, which bypasses the process of absorption, a chemical is absorbed when it crosses the epithelial layers in the skin, small intestine, or alveoli in the lungs and enters the bloodstream from an external site of exposure.

Compared to the epithelium in the small intestine, the skin is relatively impermeable to aqueous solutions and ions, but it may be permeable in varying degrees to a large number of drugs or xenobiotics. Drug or xenobiotic delivery pathways in the skin and GI tract can hypothetically involve intercellular and intracellular passive diffusion across the epidermis and transappendageal routes via hair follicles and sweat pores in the skin. Transappendageal pathways are considered to contribute very little to the dermal transport of most drugs compared to transport across the epidermis (Barry, 1991). It is possible for very small and/or polar molecules to penetrate through these appendages or shunts, but very unlikely for many classes of highly lipophilic pesticides. The stratum corneum cell layer in human skin (10–50 μm) and pig skin (15 μm) is nonviable and is considered to be the rate-limiting barrier in percutaneous absorption of many drugs and pesticides (Monteiro-Riviere et al., 1990). Most available research has concentrated on the stratum corneum as the primary barrier to absorption, although the viable epidermis (ca. 80 μm in humans and 60 μm in pigs) and dermis (3–5 mm in humans) may contribute significantly to the percutaneous penetration of drugs and ultimately their bioavailability. Scheuplein (1972) proposed that polar drugs diffused through the hydrated keratin of the dead cells in the stratum corneum, whereas nonpolar drugs traversed the intracellular lipid. The accepted hypothesis is that the dominant pathway for polar molecules resides in the aqueous region of the intercellular lipid with the hydrophobic region of the lipid chains providing the nonpolar route (Elias, 1981). The intercellular region, as depicted in the brick and mortar model of the stratum corneum, and now considered the most likely path for absorption of lipophilic drugs and pesticides, is filled with neutral lipids (complex hydrocarbons, free sterols, sterol esters, free fatty acids, and triglycerides), which make up 75% of the total lipids, and polar lipids, such as phosphatidylethanolamine, phosphatidylcholine, lysolecithin, ceramides, and glycolipids (Magee, 1991). Percutaneous and GI absorption through the intercellular pathway is by passive diffusion and it is often correlated to the partition coefficient. The rate of absorption of the penetrant can be described by Fick's law of diffusion.

Continuous blood flow removes the xenobiotic from the site of absorption in the skin and GI tract, thus maintaining a concentration gradient and enhancing continued absorption. For many of the lipophilic pesticides, penetrating molecules are thought to enter the systemic circulation at the dermis/epidermis interface in skin and do not necessarily traverse the full thickness of the dermis. For rapidly absorbed chemicals, equilibrium may be established between the blood and the site of absorption, and the rate of entry into the blood is limited by blood flow rather than by diffusion across the membrane. In this case, an increase in blood flow will increase the rate of absorption of the chemical and absorption is said to be perfusion (or blood-flow) limited. For poorly absorbed chemicals, however, absorption is not sensitive to blood flow and is said to be diffusion-rate limited.

Percutaneous Absorption

The skin is a complex tissue with a large surface area, whose primary functions are to protect the body from physical or chemical insult, to thermoregulate, and to simultaneously prevent water loss from the body. Dermal absorption of any chemical requires movement from the environment across this barrier, which is a biochemical milieu of complex lipids and proteins. Experimentally, there are several in vitro, ex vivo, and in vivo models that have been used to estimate dermal absorption of pesticides in humans. Although in vivo methods are the gold standard, each of these methods has its respective weaknesses and strengths for accurately predicting the dermal absorption of pesticides. Dermal absorption assessment is further complicated by species, age, and sex differences and differences between anatomical sites within a species. More importantly, dermal absorption in rodent skin is not always equivalent to that in human skin.

Dermal absorption is dependent on the physicochemical properties of the pesticide, the formulation, and the environmental conditions. The pesticide applicator is often clothed and operating in extreme environments, not standard laboratory conditions. This section of the chapter focuses on the differences in absorption between anatomical body sites and the effects of formulation chemistry and environmental factors that influence percutaneous absorption of pesticides.

Anatomical Site Differences

Regional variation in skin permeability at different body sites may be related to skin thickness, number of cell layers, cell size of the epidermis and stratum corneum, and distribution of hair follicles and sweat pores. Because of thick layers of stratum corneum, permeability in palmar and plantar skin is expected to be less than that in the scalp or forearm (Feldmann and Maibach, 1974). Data from several studies suggest that regional variation in vascular anatomy and blood flow should also be considered (Monteiro-Riviere et al., 1990; Qiao et al., 1993).

Various studies have demonstrated regional variation in penetration of drugs and pesticides in pig skin (Qiao and Riviere, 1995; Qiao et al., 1993), rat skin (Bronaugh, 1985), and rhesus monkey skin (Wester et al., 1980). These studies further demonstrated that parathion penetrates nonoccluded pig skin in the decreasing order of back, shoulder, buttocks, abdomen; for occluded skin, the order is back, abdomen, buttocks, shoulder. Wester et al. (1994) also demonstrated that pyrethrin absorption through the human forearm is less than the predicted absorption in the human scalp. This anatomical difference is somewhat consistent with lindane absorption through the forearm (18%), forehead (34%), and palm (34%) of rhesus monkeys (Moody and Ritter, 1989). This anatomical range for lindane is similar to that for dermal absorption of DEET (diethyl-*m*-toluamide) in rhesus monkeys (Moody et al., 1989). There are also data to suggest that dermal absorption of permethrin, aminocarb, DEET, and fenitrothion in monkey foreheads is twice that in monkey forearms (Moody and Franklin, 1987; Moody et al., 1987; Sidon et al., 1988).

However, Moody et al. (1990, 1992) demonstrated that there is no difference between the absorption of the acid and the amine forms of 2,4-dichlorophenoxyacetic acid (2,4-D) in rhesus monkey forearm and forehead and forearm and palm regions. The palmar absorption data conflict with the accepted dogma that absorption through palmar skin should theoretically be less than that through forearm skin because of the thickness of the stratum corneum in palmar skin (Maibach et al., 1971). It is proposed that because of the hydrophilic nature of 2,4-D-amine, absorption can occur through polar routes such as eccrine glands, which are more frequent in the palmar skin than in the forearm skin. This anatomical difference does not explain the discrepancy with lindane, which is more lipophilic than 2,4-D and least likely to be absorbed via a polar route.

Despite a 3-fold range in follicle area in the marmoset, no differences in absorption rates of paraquat, mannitol, water, and ethanol were observed between various body sites (Scott et al., 1991). However, among the various species examined in this study, there was an 80-fold range in follicle area, which correlated with observed differences in the rates of mannitol and paraquat absorption. The authors concluded that this correlation was possible only with relatively slowly absorbed test penetrants such as paraquat and mannitol. Further work is needed to determine the extent to which the unique anatomical features at different body sites play a role in the absorption and penetration of both lipophilic and hydrophilic pesticides.

Pesticide Formulation and Mixtures

Insecticide efficacy, the stability of active ingredients, and programmed release of active ingredients from the vehicle/device are the most important characteristics controlled for when pesticides are formulated (Krenek and Rohde, 1988). Environmental Protection Agency (EPA) registration does not always require percutaneous absorption studies. For this reason, more efficacy data than dermal pharmacokinetic data are available in the literature. Furthermore, most of the available pesticide absorption data pertain to binary mixtures (pesticide + vehicle). Technical grade formulations are, however, complex mixtures of formulation additives and, therefore, risk assessment based on data from exposure to binary mixtures may be inappropriate. Pesticides are usually formulated to contain active and inactive or inert ingredients. The latter can enhance the rate and extent of absorption or slow the release of the active ingredient and thus reduce the rate and extent of absorption (Walters and Roberts, 1993). These "inert" ingredients are often classified as adjuvants, surfactants, preservatives, solvents, diluents, thickeners, and stabilizers. These pesticide additives were first covered by the Food and Drug Administration and now are covered by EPA regulation 40 CFR 180.1001 and also TSCA and FIFRA (Seaman, 1990). This increasing list of inerts as well as the prohibitive cost of obtaining 40 CFR 180.1001 clearance of new inerts strongly supports the need to evaluate the influence of current and novel inerts on the toxicology and dermal absorption of active ingredients in pesticide formulations.

Several studies have demonstrated the penetration-enhancing ability of acetone compared to water, ethanol, or other vehicles commonly used in dermal absorption studies. Early work by O'Brien and Dannelley (1965) showed that in comparison with benzene and corn oil, acetone was best at enhancing carbaryl absorption. More recent studies have also demonstrated the enhancing effect of acetone compared with other solvent systems on the absorption of carbaryl, p-nitrophenol, and 2,4-D (Baynes and Riviere, 1998; Brooks and Riviere, 1995; Moody et al., 1992).

However, other studies have demonstrated that commercial formulations are more effective than acetone at enhancing pesticide absorption. Methyl parathion absorption in vitro in human skin at 24 h was 1.3% in acetone, but was significantly increased to 5.2% in a commercial formulation (Sartorelli et al., 1997). Likewise, in vivo dermal exposure studies of lindane in humans resulted in approximately 60% with a white spirit formulation and 5% with an acetone vehicle (Dick et al., 1997a,b). In these experiments, more of the lindane dose (79%) remained on the skin surface at 6 h with acetone than with the white spirit formulation (10.5%), and significant levels of lindane accumulated in the stratum corneum with white spirit (30%) and with acetone (14.3%) at 6 h. These findings strongly suggest that the white spirit formulation enhanced lindane penetration with respect to the acetone vehicle. The in vitro studies with human skin also demonstrated a similar pattern, although only 18 and 0.3% of the dose was absorbed into the perfusate at 6 h using the white spirit formulation and the acetone vehicle, respectively. Topical application of 1% commercial lotion of lindane in vitro in human and guinea pig skin resulted in absorption levels as high as 71.72 and 35.31%, respectively, at 48-h exposure (Franz et al., 1996).

Dermal absorption of alachlor as an emulsifiable concentrate and microencapsulated formulation was demonstrated to be 8.5 and 3.8%, respectively, in rhesus monkeys after a 12-h exposure (Kronenberg et al., 1988). About 88% of the systemically absorbed dose was excreted in urine within 48 h. However, the differences between these two formulations were not statistically significant. Although dilution of either of these formulations (1:29) slightly enhanced alachlor absorption, these effects were surprisingly not statistically significant. One in vitro study with human skin demonstrated similar absorption data (0.5–4%) after an 8-h exposure and peak fluxes within 3–5 h postapplication (Bucks et al., 1989b). However, a significant effect of formulation dilution with water was observed in this study, even though the same mass of alachlor was applied to skin. Not surprisingly, a greater fraction of alachlor was present on the skin surface and skin tissue than in the receptor fluid, and the high capacity for stratum corneum binding demonstrated in this study is not unique for related chlorinated aromatic chemicals.

Data from several studies have demonstrated that pesticide applicators may be at risk of increased dermal absorption of some pesticides if they apply sunscreen or an insect repellent while working with pesticides. The active ingredients in many commercial sunscreens contain UV absorbers (e.g., titanium dioxide and zinc oxide), which could act as penetration enhancers of 2,4-D, paraquat, parathion, malathion (Brand et al., 2002, 2003, 2007). Other studies have demonstrated that the active ingredient in many insect repellents, DEET, enhances transdermal delivery of drugs and toxicants (Moody et al., 1987; Windheuser et al., 1982). Some studies have demonstrated that DEET can act as a transdermal accelerant of 2,4-D-amine (Moody et al., 1992). Recent studies in our laboratory have, however, determined that DEET blocked permethrin absorption and inhibited carbaryl absorption in acetone, but not in dimethyl sulfoxide mixtures (Baynes and Riviere, 1998; Baynes et al., 1997). The insecticide synergist piperonyl butoxide, which is often formulated with some insecticidal products, was also shown to enhance carbaryl absorption (Baynes and Riviere, 1998). These diffusion studies further demonstrated that piperonyl butoxide does not enhance absorption through inert latex membranes, but does so in porcine skin sections. This observation suggests that some chemical-biological interaction or other mechanisms (e.g., irritation) may occur in skin to enhance the absorption of pesticides. An expected, but important finding in these carbaryl experiments was that increased dilution of the carbaryl formulation with water, especially in the presence of the surfactant sodium lauryl sulfate (SLS), enhanced carbaryl absorption. The penetration-enhancing effect of SLS was also observed with parathion (Qiao et al., 1996).

In addition to the formulation additives, agrochemicals may contain isomers, homologs, or breakdown products that form after synthesis and/or formulation and during storage (Chambers and Dorough, 1994). Although these impurities can potentially alter the toxicity and toxicokinetics of the pesticide, many toxicology and dermal absorption studies have ignored these impurities and used the pure rather than the technical-grade pesticide. There is evidence that technical-grade malathion can be more lethal (eight-fold difference) in rats than the purified form (Umetsu et al., 1977). Other studies have demonstrated that organophosphates such as malathion and fenitrothion can potentiate the toxicity of the carbamate insecticide carbaryl (Takahashi et al., 1987).

Previous metabolism studies in isolated perfused porcine skin flaps (Carver et al., 1990) demonstrated a significant first-pass metabolism of parathion to p-nitrophenol and paraoxon, and that these metabolites may be present simultaneously during absorption of parathion. Environmental exposure to parathion is never to pure parathion because spontaneous degradation occurs during storage. When mixtures of parathion and its metabolites were dosed and then assayed for parathion and its two metabolites across pig skin in vitro, significant interactions were detected. In general, the nontoxic metabolites p-nitrophenol and 1-naphthol can significantly enhance the absorption of the parent compounds, parathion and carbaryl, respectively (Baynes and Riviere, 1998; Chang et al., 1994). Surprisingly, p-nitrophenol did not enhance the absorption of paraoxon; this toxic metabolite of parathion appears to decrease the absorption of *p*-nitrophenol and paraoxon. In other related absorption studies, pretreatment with 3% fenvalerate decreased subsequent absorption of parathion, increased subsequent lindane absorption, and had no effect on subsequent fenvalerate or carbaryl absorption (Chang et al., 1995). These results underscore the chemical specificity of these interactions and reinforce the concept that the percutaneous absorption of a mixture cannot be predicted from individual component studies.

These data suggest that other mechanisms in addition to vehicle and surfactant effects must be operating simultaneously; hence further investigation is required. The data reinforce the concept that the permeability of a mixture cannot be predicted from individual component studies. Many of the mechanisms of pesticide mixture interactions are not well understood and are not easy to model, although a biophysical model for parathion has been attempted (Williams et al., 1996). It should also be recognized that it is more often the formulation additives and other environmental factors rather than the active ingredient that compromise the skin barrier and eventually enhance pesticide absorption. There is epidemiological evidence that agricultural pesticides can cause dermatoses (Abrams et al., 1991; Cellini and Offidani, 1994; Guo et al., 1996), and there is experimental evidence that UV irradiation can enhance skin reactions to topical agricultural chemical treatment (Kimura et al., 1998). In the latter study, significant reactions were observed for several herbicides. Maibach and Feldmann (1974) demonstrated that dermal absorption of pesticides such as parathion, azodrin, and diquat occurs more readily (ninefold) through damaged skin than through normal skin. It is, therefore, plausible to assume that the formulation additive can inflict local reversible or irreversible damage to the skin structure and physiology and that it is these interactions that modulate dermal absorption of most pesticides.

Recent in vivo animal studies have demonstrated that oral consumption of alcohol can significantly increase the dermal absorption of the herbicides 2,4-D, paraquat, and atrazine and the insect repellent DEET by as much as 1.6- to 2.3-fold (Brand et al., 2004, 2007). The authors of these studies proposed that alcohol solvates the polar head regions of the lipid in the stratum corneum and thus disrupts the interactions between the polar head group and alkyl chains. Further work by these investigators demonstrated that oral ingestion of alcohol significantly enhanced skin lipid peroxidation and transepidermal water loss and this is a more plausible explanation for the increased dermal absorption of these pesticides (Brand and Jendrzejewski, 2008).

Environmental Factors
(a) Temperature
Changes in ambient air temperature can alter lipid fluidity in the intercellular lipid domain of the stratum corneum. This alteration in the intercellular pathway can theoretically alter pesticide penetration through the stratum corneum. Previous in vivo studies have demonstrated that increased percutaneous absorption of a cholinesterase

inhibitor (VX) was a function of skin temperature (Craig et al., 1977). In humans topically exposed to parathion at various ambient temperatures (11, 25, and 40°C), the urinary excretion of the metabolite p-nitrophenol paralleled the increase in ambient temperature (Hayes et al., 1964). Several in vitro experiments with pig skin also demonstrated that increasing the air temperature from 37 to 42°C significantly increased parathion absorption (Chang and Riviere, 1991).

Increased ambient temperatures can also increase the evaporation of volatile pesticides from the skin, thereby reducing the topical dose available for absorption. Increasing air flow over the skin increases evaporative loss and significantly decreases dermal residues in the upper skin layer of pigs for DDT (dichlorodiphenyltrichloroethane), malathion, parathion, and DEET (Reifenrath et al., 1991). Wester et al. (1992a) demonstrated that isophenfos concentrations on the human skin surface in vivo were less than 1% dose at 24 h and that evaporation from the skin surface during absorption reduced the dose available for penetration and absorption. Finally, it should be recognized that skin surface conditions in vitro are more easily controlled than in vivo, and data from in vitro studies can significantly underestimate evaporation in vivo.

(b) Humidity and Occlusion

Skin hydration can be increased by occlusion, with high relative humidity or immersion conditions (e.g., swimming or bathing). Although previously it was assumed that hydration changes affect dermal absorption of only polar compounds, there are significant data that suggest that at high relative humidity, this hydration effect becomes more important for nonpolar molecules such as pesticides, and is most probably secondary to an increase in diffusivity of the penetrating molecule (Behl et al., 1980). Under relative humidity conditions greater than 80%, parathion absorption was significantly increased in pig skin in vitro by as much as two to three times the value under standard conditions of 60% relative humidity (Chang and Riviere, 1991, 1993).

The practical application of occlusion is when pesticides get into and under the clothing of workers; this creates the ideal reservoir for penetration and absorption into the skin. Occlusion can change dermal absorption by various mechanisms, such as reduction in evaporation from the skin surface, enhancement of skin hydration, changes in cutaneous metabolism, dermal irritation, and altered cutaneous blood circulation (e.g., vasodilation). Occlusion can increase hydration of the stratum corneum from as little as 5–15% to as much as 50% (Bucks et al., 1989a), thereby modulating the absorption profile for the pesticide. One in vivo study with pigs (Qiao et al., 1997) demonstrated that occlusion significantly enhanced pentachlorophenol (PCP) absorption from 29.1 to 100.72% dose and changed the shape of the absorption profile in blood and plasma. The study also suggested that occlusion changed the local metabolism of PCP and, as a result, the ^{14}C partitioning between plasma and red blood cells. Occlusion was also kinetically related to modification of cutaneous biotransformation of topical parathion (Qiao and Riviere, 1995). Occlusion enhanced the cutaneous metabolism of parathion to paraoxon and to *p*-nitrophenol as well as the percutaneous absorption and penetration of both parathion and *p*-nitrophenol. Occlusion also reduced parathion and *p*-nitrophenol levels in the skin, but increased *p*-nitrophenol and *p*-nitrophenol glucuronide in the blood.

Other in vivo studies (Qiao et al., 1993) showed that dermal occlusion significantly enhanced the rate and extent of parathion absorption in pigs in the abdomen (43.94 vs 7.47%), buttocks (48.47 vs 15.60%), back (48.82 vs 25.00%), and shoulder (29.28 vs 17.41%). Although significant anatomical site differences were observed with nonoccluded skin, these site differences were concealed with occluded skin. In vitro studies with parathion also demonstrated that occlusion increased absorption from 0.46–7.69 to 1.04–17.46% at doses ranging from 4 to 400 μg/cm^2 (Chang and Riviere, 1993).

Pesticides can be transferred from cotton fabric onto and through human skin as demonstrated in several studies (Snodgrass, 1992; Wester et al., 1996b), but it should be recognized that these studies were under occlusive conditions. Dermal absorption of malathion was 3.92% with ethanol-wet fabric and 0.6% with 2-day-treated cotton sheets (Wester et al., 1996b). However, malathion absorption was increased to 7.34% when the 2-day-treated/dried cotton fabric was wetted with aqueous ethanol. In the same study, absorption of glyphosate was 1.42% in water solution, 0.74% when applied as wet cotton sheets, and 0.08% when applied as 2-day-treated/dried cotton sheets. Absorption increased to 0.36% when the 2-day-treated/dried cotton sheets were wetted with water to simulate sweating and wet conditions. Military uniforms are impregnated with permethrin as a defense against nuisance and disease-bearing insects. Application of fabric impregnated with permethrin to the backs of rabbits resulted in a 3.2% migration to the skin surface with 2% of the impregnant being absorbed and 1.2% remaining on the skin surface after 7 days of continuous skin contact (Snodgrass, 1992). The implications of these interactions, especially for agricultural workers during pesticide application in humid climates or for military personnel under combat conditions in the desert, should not be underestimated.

(c) Soil

Pesticide adsorption to soil can alter the amount of pesticide available for dermal absorption. It should also be recognized that exposure conditions such as exposure time, pesticide concentration, soil load, and soil characteristics are important variables that can theoretically influence absorption (Bunge and Parks, 1997). Soil adherence to skin, for instance, can vary from 10^{-3} to 10^2 mg/cm^2 and has been shown to be activity-dependent (Kissel et al., 1996). Predicting dermal absorption of pesticides from contaminated soils is, therefore, not a simple process and becomes problematic because there are very few studies that have addressed many of these issues. For several pesticides (e.g., PCP, 2,4-D, chlordane), percutaneous absorption in an acetone vehicle appears to be slightly less or not significantly different from absorption from soil. However, for several other pesticides (e.g., DDT, organic arsenicals), soil appears to reduce percutaneous absorption of the pesticide. The interactions between soil and several of these pesticides are subsequently described in more detail, but note that in vitro skin models are, in general, not very predictive of in vivo absorption when exploring these interactions (Wester and Maibach, 1998).

Although DDT is no longer widely used in the United States, residues in soil are still detectable and human contact with contaminated soil can result in DDT exposure. One study demonstrated that in vivo absorption of DDT in rhesus monkeys was significantly less from soil (3.3% dose) than from acetone vehicle (18.9%) (Wester et al., 1990). The absorption of DDT in acetone in rhesus monkey is not significantly different from DDT absorption in humans (10.4% dose; Feldmann and Maibach, 1974). In vivo absorption from acetone or soil was not similar to in vitro absorption (0.1%). However, in vitro experiments demonstrated that 18.1% penetrated skin with acetone and 1.0% penetrated skin with soil. Less than 1.0% dose partitioned into the receptor phase, demonstrating that the skin barrier in addition to the soil is rate limiting and that in vitro skin models may not be useful for predicting DDT absorption in vivo.

Unfortunately, only in vitro dermal absorption studies are available for organic arsenicals. One study demonstrated that as much as 12.4% dose of monosodium methyl arsenate (MSMA) and disodium methyl arsenate (DSMA) penetrated mouse skin within 24h from aqueous vehicles over a wide dosage range (Rahman and Hughes, 1994). Of this amount, only 4% was absorbed into the receptor fluid. In the presence of soil (690 ppm), penetration through mouse skin was reduced to not more than 0.48 and 0.22% for MSMA and DSMA, respectively. Increasing MSMA and DSMA levels in soil from 690 to 6900 ppm increased skin content, but decreased the percentage of applied dose in skin. Whereas absorption into receptor fluid was very low for MSMA (0.01%), it was not detectable for DSMA. Topical application of aqueous solutions (20, 100, and 250 μl) of 10 μg of dimethylarsenic acid (DMA) to mouse skin resulted in 5.16–25.22% dose in receptor fluid and 1.95–15.67% dose in skin tissue within 24h (Hughes et al., 1995). However, when DMA (690 ppm) was applied with soil, absorption was reduced to 0.08% in the receptor fluid and 0.45% in skin.

The influence of soil was, however, not observed with inorganic arsenic. In vivo percutaneous absorption of arsenic as $H_3A_5O_4$ in water in rhesus monkeys (2.0–6.4%) was somewhat comparable to in vitro absorption (1.9%) in human skin (Wester et al., 1993a). However, the soil vehicle did not influence absorption in rhesus monkeys (3.2–4.5%) or human skin in vitro (0.8%), although absorption in these skin models is not comparable. The relative similarities in the partition coefficient of arsenic from water to stratum corneum and from water to soil probably explain why absorption from water was similar to absorption from soil.

Interactions between soil and the phenoxy herbicides (e.g., 2,4-D acid, 2,4-D amine) are unique. One study demonstrated that dermal absorption of the herbicide 2,4-D acid is nonlinear with respect to soil load or skin contact time (Wester et al., 1996a). Percutaneous absorption in an acetone vehicle (8.6%) was not different from absorption of soil loads of $1 \, mg/cm^2$ (8.6%) and $40 \, mg/cm^2$ (15.9%) in rhesus monkey in vivo. Further in vitro experiments with human skin demonstrated that increasing the soil load from 5 to $40 \, mg/cm^2$ did not affect 2,4-D absorption, which ranged from 1.4 to 1.8%. During the first 24 h of in vitro exposure, absorption was linear with respect to time for an acetone vehicle (3.2%); however, there was an apparent lag time of about 8 h with absorption from a soil vehicle (0.03–0.05%). This early lag time may be related to chemical partitioning from soil and may be beneficial if the skin is decontaminated within 24 h. The investigators proposed that because of complex interactive forces between pesticides and soil, dermal absorption calculations based on assumed linearity can incorrectly estimate the threat to human health. Mathematical extrapolation from high soil loads to low soil loads may significantly underestimate 2,4-D absorption. These studies also demonstrated that soil release kinetics may limit dermal absorption and that more data are needed to make valid predictions. It is therefore plausible to assume that only pesticides in the soil layer that is in direct contact with skin are bioavailable and heavy soil loads may not necessarily increase dermal absorption.

In contrast to DDT, chlordane absorption in rhesus monkeys in acetone (6.0% dose) was similar to absorption in soil (4.2% dose) 6 days after exposure (Wester et al., 1992b). Although human skin in vitro experiments demonstrated similar partitioning into receptor fluid for acetone (0.07%) and soil vehicles (0.04%), there was greater penetration into skin with acetone (10.8%) than with soil (0.34% dose) at 24 h. It is possible that chlordane adsorption to soil delayed percutaneous absorption during the initial 24 h and an extrapolation to 6 days would reveal no vehicle differences as demonstrated in the in vivo study. The octanol:water partitioning coefficients (log P) of chlordane and DDT are 5.58 and 6.91, respectively, and, therefore, dermal disposition should be similar. The high lipophilicity of these pesticides explains the higher proportion of pesticide in the skin than in the receptor phase, but it does not explain why the differences between acetone and soil for DDT are greater than those for chlordane; it only suggests that factors other than lipophilicity influence absorption of these organochlorines.

Various studies have demonstrated that the very ubiquitous pesticide, PCP, is very readily absorbed through human, monkey, and pig skin (Qiao et al., 1997; Wester et al., 1993b). In vivo absorption of PCP in rhesus monkeys with acetone vehicle (29.2%) was similar to absorption with soil vehicle (24.4%) after a 24-h exposure period (Wester et al., 1993b). However, in vitro absorption in human skin appears to underestimate in vivo absorption because only 0.6–1.5 and 0.01–0.07% dose were detected in receptor fluid with acetone and soil vehicle, respectively, at 24 h. Skin concentrations were only 2.6–3.7 and 0.11–0.14% for acetone and soil vehicles, respectively, which is still not comparable to the in vivo data. The approximately 25% PCP absorption from nonocclusive soil in monkeys (Wester et al., 1993b) compares favorably with the 29% PCP absorption from nonocclusive soil in Yorkshire pigs in vivo (Qiao et al., 1997).

Note that inhibition of soil and/or skin microorganisms can inhibit absorption of PCP, alter local and systemic distribution, and increase plasma/blood concentration ratios in pig skin in vivo (Qiao et al., 1997). It is plausible to assume that skin or soil microorganisms and/or products of PCP microbial degradation may play a role in PCP absorption and disposition.

Absorption from the Gastrointestinal Tract

People are potentially exposed to pesticides orally from pesticide residues in foods such as meat, milk, fruits, and vegetables. Children may also be orally exposed to pesticides when they place contaminated objects in their mouths. The rate and extent of absorption after oral exposure depend on the ability of the chemical to cross the plasma membranes of the gastrointestinal tract. As discussed under Factors That Influence the Transfer and Availability of Chemicals in the Body, diffusion across a plasma membrane depends to a large extent on the lipid solubility and degree of ionization of the chemical. The degree of ionization of weak acids and weak bases, and hence absorption, depends on pH. The pH range in the gastrointestinal tract varies from approximately 1–3 in the stomach to 6–8 in the intestines. Thus, the rate and extent of absorption of weak organic acids and bases vary with location in the gastrointestinal tract; weak acids are nonionized and are absorbed in the stomach, whereas weak bases are nonionized and are absorbed in the intestine (Figure 3.4). Removal from the site of absorption by blood flow maintains a concentration gradient, thus enhancing the absorption of chemicals.

Residence time in the region of the gastrointestinal tract where the chemical is absorbed also affects absorption. The presence of food in the gut can alter the pH of the gut contents and the intestinal motility, which in turn can affect the rate of absorption from the gastrointestinal tract. Stomach acid, gastric enzymes, and intestinal flora may decompose the chemical before absorption can occur, which may also decrease the potential for toxicity. Pekas (1972) demonstrated the hydrolysis of naphthyl N-methyl carbamate in the intestine (pH 6.4) and Baynes and Bowen (1995) demonstrated the effect of rumen microflora on methyl parathion. The large surface area of the intestinal tract aids in absorption from this site; even chemicals that do not readily cross the plasma membrane (e.g., weak acids) can be absorbed to a high degree in the intestine because of the increased surface area. Particles may be absorbed in the intestines by endocytosis.

Chemicals that are absorbed into the bloodstream from the gastrointestinal tract enter the portal circulation and are delivered directly to the liver, where they may

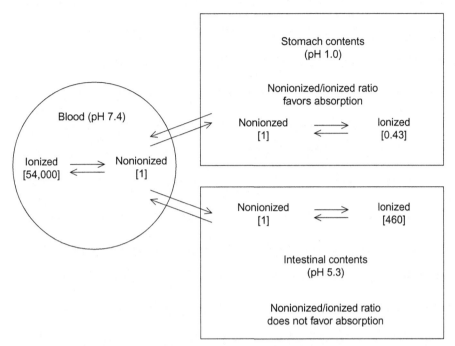

Figure 3.4 Proportion of nonionized and ionized forms of 2,4-D ($pK_a < 2.6$) in the stomach and intestinal contents *(adapted from Hodgson et al., 1991)*. Only the nonionized form crosses cell membranes.

be metabolized before reaching the systemic circulation. This first-pass metabolism decreases the systemic availability (bioavailability) of the parent compound. Some highly lipophilic compounds such as organochlorine pesticides are absorbed into the lymphatic system in a manner similar to the absorption of nutritional fats (Turner and Shanks, 1980). Absorption into the lymphatic system bypasses delivery to the liver and the potential for first-pass metabolism.

Many of the in vivo oral absorption studies in laboratory animals have demonstrated that oral absorption of many pesticides can be significantly greater than absorption following skin exposure. This is especially applicable to the pyrethroid class of insecticides, which do not readily penetrate skin and usually have a dermal bioavailability of less than 10% in human skin. Deltamethrin is rapidly absorbed ($T_{max} = 1.0\,h$) and its absolute oral bioavailability was 18% in rats given doses in the range of 0.2–10 mg/kg by oral gavage in glycerol formal (Kim et al., 2007). Permethrin was less rapidly absorbed ($T_{max} = 3.52\,h$) and had a greater absolute bioavailability ($F = 61\%$) in the same strain of rats (Anadon et al., 1991). Many of the organophosphates are readily absorbed, although the data suggest that pesticide formulation can result in absorption varying from 20% to as much as 70% according to several human studies (Eaton et al., 2008;

Nolan et al., 1984; Timchalk et al., 2002). The commonly used chlorophenoxy herbicide 2,4-D is rapidly absorbed by the human gastrointestinal tract (Kohli et al., 1974; Sauerhoff et al., 1977); however, a more recent rodent study suggests that clearance can be uniquely sex-dependent (Griffin et al., 1997).

The quaternary nitrogen herbicide paraquat has a limited bioavailability in laboratory animals of about 22% and less than 5% absorption during the first 6 h (Chui et al., 1988; Meredith and Vale, 1987), although it has often been associated with human poisonings following accidental or suicidal ingestion. Paraquat absorption across the GI tract can be rapid, with the peak plasma concentration occurring within the first hour after exposure (Heylings, 1991; Nagao et al., 1993) and it is plausible to assume that the presence of herbicide performance surfactants may play a significant role in paraquat uptake across the GI tract. There have been recent successful efforts to include additives such as alginate in paraquat formulations to reduce paraquat uptake by the GI tract of experimental animals. The mechanism involves delaying gastric emptying, which prevents early high lethal dose delivery to the lungs and thus enhances human survival following acute exposure to this lethal herbicide (Heylings et al., 2007).

Atrazine, which belongs to the class of triazine herbicides, displays unique absorption kinetics that are not often observed with many pesticides. While one rodent study has estimated that oral absorption may be slow, with an absorption half-life of 3 h, and limited to approximately 57% based on 24-h urine excretion data (Timchalk et al., 1990), more recent rodent studies demonstrated a double-peak phenomenon as evidenced by a single first-order absorption process and a longer plateau (McMullin et al., 2003, 2007). These studies also estimated that limited solubility in the intestine and presystemically metabolized atrazine in the intestine may account for limited oral bioavailability. However, it should be emphasized again that oral absorption is significantly greater through oral exposure than by dermal exposure, as exemplified here by atrazine, for which only as much as 5.6% of the topical dose is expected to be absorbed across the skin of humans and as much as 10 times this amount could be absorbed by the gastrointestinal tract. Risk assessors should be wary of formulation and dose effects, which will alter bioavailability across both routes of exposure.

Absorption from the Respiratory Tract

A chemical must be in the form of a gas, vapor, or particulate (e.g., aerosol) to be absorbed in the respiratory tract. Although the anatomy of the respiratory tract varies widely within mammalian species, the respiratory system can be generally compartmentalized into the nasopharyngeal, tracheobronchial, and pulmonary regions (Kennedy and Valentine, 1994). The function of the nasopharyngeal region is to condition inspired air and to remove large inspired particles before they reach the tracheobronchial and pulmonary regions. The tracheobronchial region is lined by mucus-secreting and ciliated cells, which together make up the mucociliary escalator.

The pulmonary region of the respiratory tract is the gas–exchange region, which consists of the respiratory bronchioles, alveolar ducts, and alveoli.

Inspired gases and vapors may be absorbed throughout the respiratory tract, depending on their physicochemical properties, and the anatomy and physiology of the region. Inhaled gases and vapors diffuse across cell membranes in the direction of the concentration gradient until equilibrium is established (see also Transport Mechanisms). The ratio of the gas or vapor equilibrium concentrations in blood and air is termed the blood:air partition coefficient. Highly water-soluble and reactive gases and vapors tend to be absorbed in the mucus layer of the upper respiratory tract, whereas more lipophilic and nonreactive gases and vapors are absorbed from the deeper regions of the respiratory tract. The geometry, blood flow, and capacity for metabolism of the respiratory tract may also influence the rate and site of absorption of inhaled gases and vapors.

Deposition of aerosols in the respiratory tract depends on a number of factors, including the physicochemical and aerodynamic properties of the aerosol and the geometry and airflow of the respiratory tract. Airflow (velocity) and turbulence decrease from the nasopharyngeal to the pulmonary region, and different mechanisms of deposition operate in these regions. Impaction is an important mechanism for deposition of particles larger than $1\,\mu m$ in aerodynamic diameter in regions of the respiratory tract where air velocity and turbulence are high, such as airway bifurcations. Interception is an important mechanism of deposition for fibers and is dependent on fiber length rather than diameter. Sedimentation is settling due to gravity and is important for particles larger than $1\,\mu m$ in aerodynamic diameter in regions of the respiratory tract where airways are small in diameter and airflow is low. Diffusion is an important mechanism of deposition for small particles throughout the respiratory tract, and particularly for particles $<0.5\,mm$ in the alveolar region where airflow is low.

Particles are cleared from the respiratory tract in a number of ways, depending on the region of the respiratory tract. In the nasopharyngeal region, particles are removed by nose wiping, nose blowing, sneezing, and swallowing. In the tracheobronchial region, particles are cleared by the mucociliary escalator and are ultimately swallowed or expectorated. In the pulmonary region, particles are removed by (1) dissolution and removal in the bloodstream or lymphatics, (2) alveolar macrophage phagocytosis and removal by the mucociliary escalator or lymphatics, and (3) direct penetration of epithelial membranes and absorption into tissue or blood.

Absorption after Exposure by Other Routes

Intravenous (iv) administration, in which the chemical is introduced directly into the blood, by definition, bypasses the process of absorption. The advantages of administering a chemical by this route include rapid achievement of effective blood concentrations, precise knowledge of the delivered dose, and the ability to deliver a chemical that would cause irritation by other routes. As will be discussed in the pharmacokinetics section of this book, the plasma concentration-time data obtained from iv administration are necessary for estimating the absolute bioavailability of a substance given by an extravascular route. The major disadvantages of iv administration are that the administered dose cannot be removed, systemic toxicity may occur with some pesticides because of the transiently high concentration achieved, sterile pesticide preparations are required, and pesticides that are insoluble in plasma may precipitate and cause embolus formation. For experimental pharmacokinetic studies, pesticides can be administered intravenously by bolus injection or infusion or added to fluid drip bags. Intra-arterial injection is very similar to iv administration, but more hazardous since the pesticide is delivered directly to the arterial circulation. This mode is often used in pharmacology to treat localized tumors that have accessible arteries for injection.

Other routes of exposure that are used in the laboratory include intraperitoneal (ip) injection, where chemicals are absorbed primarily through the portal circulation, and intramuscular or subcutaneous injection. Chemicals administered intramuscularly (im) or subcutaneously (sc) are absorbed more slowly than by ip injection. Injection is into the muscle mass, which is well perfused by the vascular system, and can result in usually rapid absorption. Lipid solubility of the administered drug is not important as both hydrophilic and charged chemicals are easily absorbed in the rich capillary networks of muscle tissue. Absorption may be modulated by the vehicle used to inject the drug into the muscle. In this case, release of drug from the vehicle becomes the rate-limiting step. The sc mode of administration is very similar to im injection except that the drug is deposited into the rich capillary beds perfusing the skin and absorption is (in general) slower than via the im route and can be very erratic. Injection of tissue-irritating substances may cause local reactions and even skin sloughing, which alter absorption. Other routes of administration are intradermal, intramammary, and subconjunctival, which are not often associated with animal testing or human routes of exposure for pesticides.

SUMMARY AND FUTURE DIRECTIONS

Human exposure to pesticides can result in absorption by passive diffusion across the epithelial layers of the skin and/or the gastrointestinal tract. As occupational exposure via the skin is the most likely source and route for human exposure, this chapter has focused on the main physicochemical, biological, and environmental factors that could significantly influence systemic absorption following dermal exposure to examples of pesticides with diverse physicochemical properties. There are several in vitro and inert membrane models (Bronaugh and Stewart, 1984; Carver et al., 1989) that can be used experimentally to evaluate membrane transport, and there has been some success in comparing absorption across in vitro and in vivo systems (Wester et al., 1998).

These in vitro systems can be used to assess the relative influence of several formulation and biological factors that determine dermal and oral absorption of pesticides without having to use in vivo animal models or humans in the early stages of formulation development or human health risk assessments. The recent improvements in computer modeling capabilities have resulted in the development of numerous quantitative structure permeability relationships (QSPRs) that have proven to be predictive of dermal and oral permeability in humans for several solutes, including many pesticides currently in use (Baynes et al., 2008; Potts and Guy, 1992; Riviere and Brooks, 2007; Zhao et al., 2002). A very popular method, the Rule of 5, utilizes similar QSPR principles and has proven to be useful as a rapid screen for compounds that are likely to be poorly absorbed orally (Lipinski et al., 1997). This rule states that if a compound satisfies any two of the following rules, it is likely to exhibit poor intestinal absorption: (1) molecular weight >500, (2) number of hydrogen bond donors >5 (a donor being any O–H or N–H group), (3) number of hydrogen acceptors >10 (an acceptor being any O or N including those in donor groups), and (4) $C \log P > 5.0$ or $M \log P > 4.15$. There are numerous other original peer-reviewed research articles that readers are encouraged to review to get a glimpse of how these QSPR models are being developed and applied to human risk assessment. It should be noted that, in many instances, many of the data used in developing these models are often obtained from multiple sources and laboratories with diverse dosing and experimental protocols, and the statistical analyses (e.g., appropriate measures of goodness-of-fit, robustness, and predictivity) and a defined application domain may not be reported. Many of these criteria are important for valid evaluation of QSAR models that could predict pesticide absorption as new regulatory frameworks such as REACH in the European Union are implemented in various jurisdictions (Bouwman et al., 2008). Caution should therefore be exercised with the mechanistic interpretation and application of these permeability models to any given exposure scenario. The future of risk assessment of pesticides will depend heavily on the quality of these permeability models and their flexibility in predicting dermal and oral absorption in a variety of human exposure scenarios at home and at work.

REFERENCES

Abrams, K., Hogan, D. J., & Maibach, H. I. (1991). Pesticide-related dermatoses in agricultural workers. *Occup. Med., 6*, 463–492.

Alvinerie, M., Dupuy, J., Kiki-Mvouaka, S., Sutra, J. F., & Lespine, A. (2008). Ketoconazole increases the plasma levels of ivermectin in sheep. *Vet. Parasitol., 157*(1-2), 117–122.

Anadon, A., Martinez-Larranaga, M. R., Diaz, M. J., & Bringas, P. (1991). Toxicokinetics of permethrin in the rat. *Toxicol. Appl. Pharmacol., 110*, 1–8.

Barry, B. W. (1991). The LPP theory of skin penetration enhancement. In R. L. Bronough & H. I. Maibach (Eds.), *In vitro percutaneous absorption: Principles, fundamentals, and applications* (pp. 165–185). Boca Raton, FL: CRC Press.

Baynes, R. E., & Bowen, J. (1995). Toxicokinetics of methyl parathion in lactating dairy goats. *J. Agric. Food Chem.*, *43*, 1598–1604.

Baynes, R. E., & Riviere, J. E. (1998). Influence of inert ingredients in pesticide formulations on dermal absorption of carbaryl. *Am. J. Vet. Res.*, *59*, 168–175.

Baynes, R. E., Halling, K. B., & Riviere, J. E. (1997). The influence of diethyl-*m*-toluamide (DEET) on the percutaneous absorption of permethrin and carbaryl. *Toxicol. Appl. Pharmacol.*, *144*, 332–339.

Baynes, R. E., Xia, X. R., Vijay, V., & Riviere, J. E. (2008). A solvatochromatic approach to quantifying formulation effects on dermal permeability. *SAR QSAR Environ. Res.*, *19*, 615–630.

Behl, C. R., Flynn, G. L., Kurihara, T., Harper, N., Smith, H., Higuchi, W. I., et al. (1980). Hydration and percutaneous absorption. I. Influence of hydration on alkanol permeation through hairless mouse skin. *J. Invest. Dermatol.*, *75*, 346–352.

Bouwman, T., Cronin, M. T., Bessems, J. G., & van de Sandt, J. J. (2008). Improving the applicability of (Q) SARs for percutaneous penetration in regulatory risk assessment. *Hum. Exp. Toxicol.*, *27*, 269–276.

Brand, R. M., & Jendrzejewski, J. L. (2008). Chronic ethanol ingestion alters xenobiotic absorption through the skin: Potential role of oxidative stress. *Food Chem. Toxicol.*, *46*, 1940–1948.

Brand, R. M., Spalding, M., & Mueller, C. (2002). Sunscreens can increase dermal penetration of 2,4-dichlorophenoxyacetic acid. *J. Toxicol. Clin. Toxicol.*, *40*, 827–832.

Brand, R. M., Pike, J., Wilson, R. M., & Charron, A. R. (2003). Sunscreens containing physical UV blockers can increase transdermal absorption of pesticides. *Toxicol. Ind. Health*, *19*, 9–16.

Brand, R. M., Charron, A. R., Dutton, L., Gavlik, T. L., Mueller, C., Hamel, F. G., et al. (2004). Effects of chronic alcohol consumption on dermal penetration of pesticides in rats. *J. Toxicol. Environ. Health A*, *67*, 153–161.

Brand, R. M., Jendrzejewski, J. L., & Charron, A. R. (2007). Potential mechanisms by which a single drink of alcohol can increase transdermal absorption of topically applied chemicals. *Toxicology*, *235*, 141–149.

Bronaugh, R. L. (1985). *In vitro* methods for percutaneous absorption of pesticides. In R. Honeycut, G. Zweig & N. N. Ragsdale (Eds.), *Dermal exposure to pesticide use.* Washington, DC: American Chemical Society.

Bronaugh, R. L., & Stewart, R. F. (1984). Methods for in vitro percutaneous absorption studies III. Hydrophobic compounds. *J. Pharm. Sci.*, *73*(9), 1255–1258.

Brooks, J. D., & Riviere, J. E. (1995). Quantitative percutaneous absorption and cutaneous distribution of binary mixtures of phenol and *p*-nitrophenol in isolated perfused porcine skin. *Fundam. Appl. Toxicol.*, *32*, 233–243.

Brunet, J. L., Maresca, M., Fantini, J., & Belzunces, L. P. (2008). Intestinal absorption of the acetamiprid neonicotinoid by Caco-2 cells: Transepithelial transport, cellular uptake and efflux. *J. Environ. Sci. Health B*, *43*, 261–270.

Bucks, D. A. W., Maibach, H. I., & Guy, R. H. (1989a). Occlusion does not uniformly enhance penetration in vivo. In R. L. Bronaugh & H. I. Maibach (Eds.), *Percutaneous absorption.* New York: Dekker.

Bucks, D. A. W., Wester, R. C., Mobayen, M. M., Yang, D., Maibach, H. I., & Coleman, D. L. (1989b). *In vitro* percutaneous absorption and stratum corneum binding of alachlor: Effect of formulation dilution with water. *Toxicol. Appl. Pharmacol.*, *100*, 417–423.

Bunge, A. L., & Parks, J. M. (1997). Predicting dermal absorption from contact with chemically contaminated soils. In F. J. Dwyer, T. R. Doane, & M. L. Hinman (Eds.), *Environmental toxicology and risk assessment: Modelling and risk assessment* (Vol. 6). ASTM STP 1317. Philadelphia: American Society Testing Mater.

Carver, M. P., Williams, P. L., & Riviere, J. E. (1989). The isolated perfused porcine skin flap. III. Percutaneous absorption pharmacokinetics of organophosphates, steroids, benzoic acid, and caffeine. *Toxicol. Appl. Pharmacol.*, *97*, 324–337.

Carver, M. P., Levi, P. E., & Riviere, J. E. (1990). Parathion metabolism during percutaneous absorption in perfused porcine skin. *Pestic. Biochem. Physiol.*, *38*, 245–254.

Cellini, A., & Offidani, A. (1994). An epidemiological study on cutaneous diseases of agricultural workers authorized to use pesticides. *Dermatology*, *189*, 129–132.

Chambers, J. E., & Dorough, G. D. (1994). Toxicological problems associated with pesticide mixtures and pesticide impurities. In R. H. S. Yang (Ed.), *Toxicology of chemical mixtures* (pp. 135–155). San Diego: Academic Press.

Chang, S. K., & Riviere, J. E. (1991). Percutaneous absorption of parathion *in vivo* in porcine skin: Effects of dose, temperature, humidity and perfusate composition on absorptive flux. *Fundam. Appl. Toxicol.*, *17*, 494–504.

Chang, S. K., & Riviere, J. E. (1993). Effect of humidity and occlusion on the percutaneous absorption of parathion *in vitro*. *Pharm. Res.*, *10*, 152–155.

Chang, S. K., Dauterman, W. C., & Riviere, J. E. (1994). Percutaneous absorption of parathion and its metabolites, paraoxon and *p*-nitrophenol, administered alone or in combination: *In vitro* flow-through diffusion cell studies. *Pestic. Biochem. Physiol.*, *48*, 56–62.

Chang, S. K., Brooks, J. D., Monteiro-Riviere, N. A., & Riviere, J. E. (1995). Enhancing or blocking effect of fenvalerate on the subsequent percutaneous absorption of pesticides *In vitro*. *Pestic. Biochem. Physiol.*, *51*, 214–219.

Chui, Y. C., Poon, G., & Law, F. (1988). Toxicokinetics and bioavailability of paraquat in rats following different routes of administration. *Toxicol. Ind. Health*, *4*, 203–219.

Cleek, R. L., & Bunge, A. L. (1993). A new method for estimating dermal absorption from chemical exposure. 1. General approach. *Pharm. Res.*, *10*, 497–506.

Craig, F. N., Cummings, E. G., & Sim, V. M. (1977). Environmental temperature and the percutaneous absorption of cholinesterase inhibitor, VX. *J. Invest. Dermatol.*, *68*, 357–361.

Dick, I. P., Blain, P. G., & Williams, F. M. (1997a). The percutaneous absorption and skin distribution of lindane in man. I. *In vivo* studies. *Hum. Exp. Toxicol.*, *16*, 645–651.

Dick, I. P., Blain, P. G., & Williams, F. M. (1997b). The percutaneous absorption and skin distribution of lindane in man. II. *In vitro* studies. *Hum. Exp. Toxicol.*, *16*, 652–657.

Eaton, D. L., Daroff, R. B., Autrup, H., Bridges, J., Buffler, P., Costa, L. G., et al. (2008). Review of the toxicology of chlorpyrifos with an emphasis on human exposure and neurodevelopment. *Crit. Rev. Toxicol.*, *38*, 1–125.

Elias, P. M. (1981). Epidermal lipids, membranes, and keratinization. *Int. J. Dermatol.*, *20*, 1–19.

Feldmann, R. J., & Maibach, H. I. (1974). Percutaneous penetration of some pesticides and herbicides in man. *Toxicol. Appl. Pharmacol.*, *28*, 126–132.

Ferrer, A., & Cabral, R. (1993). Collective poisonings caused by pesticides: Mechanism of production–mechanism of prevention. In E. Hodgson (Ed.), *Reviews in environmental toxicology* (Vol. 5, pp. 161–201). Raleigh, NC: Toxicology Communications.

Franz, T. J., Lehman, P. A., Franz, S. F., & Guin, J. D. (1996). Comparative percutaneous absorption of lindane and permethrin. *Arch. Dermatol.*, *132*, 901–905.

Galli, C. L., & Marinovich, M. (1987). Dermal toxicity of pesticides. In L. G. Costa, C. L. Galli, & S. D. Murphy (Eds.), *Toxicology of pesticides: Experimental, clinical and regulatory perspectives* (pp. 147–169). New York: Springer-Verlag.

Garrettson, L. K., & Curley, A. (1969). Dieldrin: Studies in a poisoned child. *Arch. Environ. Health*, *19*, 814–822.

Gomez-Catalan, J., To-Figueras, J., Rodamilans, M., & Corbella, J. (1991). Transport of organochlorine residues in the rat and human blood. *Arch. Environ. Contam. Toxicol.*, *20*, 61–66.

Griffin, R. J., Godfrey, V. B., Kim, Y. C., & Burka, L. T. (1997). Sex-dependent differences in the disposition of 2,4-dichlorophenoxyacetic acid in Sprague-Dawley rats, B6C3F1 mice, and Syrian hamsters. *Drug Metab. Dispos.*, *25*, 1065–1071.

Guo, Y. L., Wang, B. J., Lee, C. C., & Wang, J. D. (1996). Prevalence of dermatoses and skin sensitization associated with use of pesticides in fruit farmers of southern Taiwan. *Occup. Environ. Med.*, *53*, 427–431.

Hayes, W. J., Funckes, A. J., & Hartwell, W. V. (1964). Dermal exposure of human volunteers to parathion. *Arch. Environ. Health*, *8*, 829–833.

Heylings, J. R. (1991). Gastrointestinal absorption of paraquat in the isolated mucosa of the rat. *Toxicol. Appl. Pharmacol.*, *107*, 482–943.

Heylings, J. R., Farnworth, M. J., Swain, C. M., Clapp, M. J., & Elliott, B. M. (2007). Identification of an alginate-based formulation of paraquat to reduce the exposure of the herbicide following oral ingestion. *Toxicology*, *241*, 1–10.

Hodgson, E., Silver, I. S., Butler, L. E., Lawton, M. P., & Levi, P. E. (1991). Metabolism. In W. J. Hayes Jr. & E. R. Laws Jr. (Eds.), *Handbook of pesticide toxicology* (pp. 107–167). San Diego: Academic Press.

Hughes, M. F., Mitchel, C. T., Edwards, B. C., & Rahman, M. S. (1995). *In vitro* percutaneous absorption of dimethylarsinic acid in mice. *J. Toxicol. Environ. Health*, *45*, 279–290.

Kennedy, G. L., Jr., & Valentine, R. (1994). Inhalation toxicology. In A. W. Hayes (Ed.), *Principles and methods of toxicology* (3rd ed., pp. 805–838). New York: Raven Press.

Kim, K. B., Anand, S. S., Kim, H. J., White, C. A., & Bruckner, J. V. (2007). Toxicokinetics and tissue distribution of deltamethrin in adult Sprague-Dawley rats. *Toxicol. Sci.*, *101*, 197–205.

Kimura, T., Kuroki, K., & Doi, K. (1998). Dermatotoxicity of agricultural chemicals in the dorsal skin of hairless dogs. *Toxicol. Pathol.*, *26*, 442–447.

Kissel, J. C., Richter, K. Y., & Fenske, R. A. (1996). Field measurement of dermal soil loading attributable to various activities: Implications for exposure assessment. *Risk Anal.*, *16*, 115–125.

Kohli, J. D., Khanna, R. N., Gupta, B. N., Dhar, M. M., Tandon, J. S., & Sircar, K. P. (1974). Absorption and excretion of 2,4-dichlorophenoxyacetic acid in man. *Xenobiotica*, *4*, 97–100.

Krenek, M. R., & Rohde, W. H. (1988). An overview—solvents for agricultural chemicals. In D. A. Hovde & D. Beestman (Eds.), *Pesticide formulations and application systems* (pp. 113–127). Ann Arbor, MI: American Society Testing Mater.

Kronenberg, J. M., Fuhremann, T. W., & Johnson, D. E. (1988). Percutaneous absorption and excretion of alachlor in rhesus monkeys. *Fundam. Appl. Toxicol.*, *10*, 664–671.

Lipinski, C. A., Lombardo, F., Dominy, B. W., & Feeney, P. J. (1997). Experimental and computational approaches to estimate solubility and permeability in drug discovery and development settings. *Adv. Drug Delivery Rev.*, *23*, 3–25.

Magee, P. (1991). Percutaneous absorption: Critical factors in transdermal transport. In F. N. Marzulli & H. I. Maibach (Eds.), *Dermatoxicology* (pp. 1–36). New York: Hemisphere Publ.

Maibach, H., & Feldmann, R. (1974). Occupational exposure to pesticides. *Report to the Federal Working Group on Pest Management for the Task Group*, pp. 120–127.

Maibach, H. I., Felman, R. J., Milby, T. H., & Serat, W. F. (1971). Regional variation in percutaneous penetration in man. *Arch. Environ. Health*, *23*, 208–211.

Martin-Bouyer, G., Khanh, N. B., Linh, P. D., Hoa, D. Q., Tuan, L. C., Tourneau, J., et al. (1983). Epidemic of haemorrhagic disease in Vietnamese infants caused by warfarin contaminated talcs. *Lancet*, *1*, 230–232.

McMullin, T. S., Brzezicki, J. M., Cranmer, B. K., Tessari, J. D., & Andersen, M. E. (2003). Pharmacokinetic modeling of disposition and time-course studies with [^{14}C]atrazine. *J. Toxicol. Environ. Health A*, *66*, 941–964.

McMullin, T. S., Hanneman, W. H., Cranmer, B. K., Tessari, J. D., & Andersen, M. E. (2007). Oral absorption and oxidative metabolism of atrazine in rats evaluated by physiological modeling approaches. *Toxicology*, *240*, 1–14.

Meredith, T. J., & Vale, J. A. (1987). Treatment of paraquat poisoning in man: Methods to prevent absorption. *Hum. Toxicol.*, *6*, 49–55.

Monteiro-Riviere, N. A., Bristol, D. G., Manning, T. O., & Riviere, J. E. (1990). Interspecies and interregional analysis of the comparative histological thickness and laser Doppler blood flow measurements at five cutaneous sites in nine species. *J. Invest. Dermatol.*, *95*, 582–586.

Moody, R. P., & Franklin, C. A. (1987). Percutaneous absorption of the insecticides fenitrothion and aminocarb in rats and monkeys. *J. Toxicol. Environ. Health*, *20*, 209–218.

Moody, R. P., & Ritter, L. (1989). Dermal absorption of the insecticide lindane (1a,2a,3b,4a,5a,6b-hexachlorocyclohexane) in rats and rhesus monkeys: Effect of anatomical site. *J. Toxicol. Environ. Health*, *28*, 161–169.

Moody, R. P., Riedel, D., Ritter, L., & Franklin, C. A. (1987). The effect of DEET (*N,N*-diethyl-*m*-toluamide) on dermal persistence and absorption of the insecticide fenitrothion in rats and monkeys. *J. Toxicol. Environ. Health*, *22*, 471–479.

Moody, R. P., Benoit, F. M., Riedel, D., & Ritter, L. (1989). Dermal absorption of the insect repellent DEET (*N,N*-diethyl-*m*-toluamide) in rats and monkeys: Effect of anatomical site and multiple exposure. *J. Toxicol. Environ. Health*, *26*, 137–147.

Moody, R. P., Franklin, C. A., Ritter, L., & Maibach, H. I. (1990). Dermal absorption of the phenoxy herbicides 2,4-D, 2,4-D amine, 2,4-D isooctyl, and 2,4,5-T in rabbits, rats, rhesus monkeys, and humans: A cross-species comparison. *J. Toxicol. Environ. Health*, *29*, 237–245.

Moody, R. P., Wester, R. C., Melendres, J. L., & Maibach, H. I. (1992). Dermal absorption of the phenoxy herbicide 2,4-D dimethylamine in humans: Effect of DEET and anatomic site. *J. Toxicol. Environ. Health*, *36*, 241–250.

Nagao, M., Saitoh, H., Zhang, W. D., Iseki, K., Yamada, Y., Takatori, T., et al. (1993). Transport characteristics of paraquat across rat intestinal brush-border membrane. *Arch. Toxicol.*, *67*, 262–267.

Nolan, R. J., Rick, D. L., Freshour, N. L., & Saunders, J. H. (1984). Chlorpyrifos: Pharmacokinetics in human volunteers. *Toxicol. Appl. Pharmacol.*, *73*, 8–15.

O'Brien, R. D., & Dannelley, C. E. (1965). Penetration of insecticides through rat skin. *J. Agric. Food Chem.*, *13*, 245–247.

Opdycke, J. C., & Menzer, R. E. (1984). Pharmacokinetics of diflubenzuron in two types of chickens. *J. Toxicol. Environ. Health*, *13*, 721–733.

Pekas, J. C. (1972). Intestinal hydrolysis, metabolism and transport of a pesticidal carbamate in pH 6.5 medium. *Toxicol. Appl. Pharmacol.*, *23*, 62–70.

Potts, R. O., & Guy, R. H. (1992). Predicting skin permeability. *Pharm. Res.*, *9*, 663–669.

Qiao, G. L., & Riviere, J. E. (1995). Significant effects of application site and occlusion on the pharmacokinetics of cutaneous penetration and biotrans-formation of parathion *in vivo* in swine. *J. Pharm. Sci.*, *84*, 425–432.

Qiao, G. L., Chang, S. K., & Riviere, J. E. (1993). Effects of anatomical site and occlusion on the percutaneous absorption and residue pattern of 2,6-[ring-^{14}C]parathion *in vivo* in pigs. *Toxicol. Appl. Pharmacol.*, *122*, 131–138.

Qiao, G. L., Brooks, J. D., Baynes, R. E., Monteiro-Riviere, N. A., Williams, P. L., & Riviere, J. E. (1996). The use of mechanistically defined chemical mixtures (MDCM) to assess component effects on the percutaneous absorption and cutaneous disposition of topically exposed chemicals. *Toxicol. Appl. Pharmacol.*, *141*, 473–486.

Qiao, G. L., Brooks, J. D., & Riviere, J. E. (1997). Pentachlorophenol dermal absorption and disposition from soil in swine: Effects of occlusion and skin microorganism inhibition. *Toxicol. Appl. Pharmacol.*, *147*, 234–246.

Rahman, M. S., & Hughes, M. F. (1994). *In vitro* percutaneous absorption of monosodium methanearsonate and disodium methanearsonate in female B6C3F1 mice. *J. Toxicol. Environ. Health*, *41*, 421–433.

Reifenrath, W. G., Hawkins, G. S., & Kurtz, M. S. (1991). Percutaneous penetration and skin retention of topically applied compounds: An *in vitro–in vivo* study. *J. Pharm. Sci.*, *80*, 526–532.

Riviere, J. E., & Brooks, J. D. (2007). Prediction of dermal absorption from complex chemical mixtures: Incorporation of vehicle effects and interactions into a QSPR framework. *SAR QSAR Environ Res.*, *18*, 31–44.

Sartorelli, P., Aprea, C., Bussani, R., Novelli, M. T., Orsi, D., & Sciarra, G. (1997). *In vitro* percutaneous penetration of methyl-parathion from a commercial formulation through the human skin. *Occup. Environ. Med.*, *54*, 524–525.

Sauerhoff, M. W., Braun, W. H., Blau, G. E., & Gehring, P. J. (1977). The fate of 2,4-dichlorophenoxyacetic acid (2,4-D) following oral administration to man. *Toxicology*, *8*, 3–11.

Scheuplein, R. J. (1972). Properties of the skin as a membrane. *Adv. Biol. Skin*, *12*, 125–152.

Scott, R. C., Corrigan, M. A., Smith, F., & Mason, H. (1991). The influence of skin structure on permeability: An intersite and interspecies comparison with hydrophilic penetrants. *J. Invest. Dermatol.*, *96*, 921–925.

Seaman, D. (1990). Trends in the formulation of pesticides. *Pestic. Sci.*, *29*, 437–449.

Sidon, E. W., Moody, R. P., & Franklin, C. A. (1988). Percutaneous absorption of *cis*- and *trans*-permethrin in rhesus monkeys and rats: Anatomic site and interspecies variation. *Toxicol. Environ. Health*, *23*, 207–216.

Smith, J. G. (1988). Paraquat poisoning by skin absorption: A review. *Hum. Toxicol.*, *7*, 15–19.

Snodgrass, H. L. (1992). Permethrin transfer from treated cloth to the skin surface: Potential for exposure in humans. *J. Toxicol. Environ. Health*, *35*, 91–105.

Sreeramulu, K., Liu, R., & Sharom, F. J. (2007). Interaction of insecticides with mammalian P-glycoprotein and their effect on its transport function. *Biochim. Biophys. Acta*, *1768*, 1750–1757.

Takahashi, H., Kato, A., Yamashita, E., Naito, Y., Tsuda, S., & Shirasu, Y. (1987). Potentiations of *N*-methylcarbamate toxicities by organophosphorous insecticides in male mice. *Fundam. Appl. Toxicol.*, *8*, 139–146.

Timchalk, C., Dryzga, M. D., Langvardt, P. W., Kastl, P. E., & Osborne, D. W. (1990). Determination of the effect of tridiphane on the pharmacokinetics of [^{14}C]-atrazine following oral administration to male Fischer 344 rats. *Toxicology*, *61*, 27–40.

Timchalk, C., Nolan, R. J., Mendrala, A. L., Dittenber, D. A., Brzak, K. A., & Mattsson, J. L. (2002). A physiologically based pharmacokinetic and pharmacodynamic (PBPK/PD) model for the organophosphate insecticide chlorpyrifos in rats and humans. *Toxicol. Sci.*, *66*, 34–53.

Turner, J. C., & Shanks, V. (1980). Absorption of some organochlorine compounds by the rat small intestine—*in vivo*. *Bull. Environ. Contam. Toxicol., 24*, 652–655.

Umetsu, N., Grose, F. H., Allahyari, R., Abu-El-Haj, S., & Fukuto, T. R. (1977). Effects of impurities on the mammalian toxicity of technical malathion and acephate. *J. Agric. Food Chem., 25*, 946–953.

Walker, M., Dugard, P. H., & Scott, R. C. (1983). *In vitro* percutaneous absorption studies: A comparison of human and laboratory species. *Hum. Toxicol., 2*, 561–568.

Walters, K. A., & Roberts, M. S. (1993). Veterinary applications of skin penetration enhancers. In K. A. Walters & J. Hadgraft (Eds.), *Pharmaceutical skin penetration enhancement*. New York: Dekker.

Wester, R. C., & Maibach, H. I. (1998). Percutaneous absorption of hazardous substances from soil and water. In M. S. Roberts & K. A. Walters (Eds.), *Dermal absorption and toxicity assessment* (pp. 697–707). New York: Dekker.

Wester, R. C., Noonan, P. K., & Maibach, H. I. (1980). Variations in percutaneous absorption of testosterone in the rhesus monkey due to anatomic site application and frequency of application. *Arch. Dermatol. Res., 267*, 229–235.

Wester, R. C., Maibach, H. I., Bucks, D. A. W., & Aufrere, M. B. (1984). *In vivo* percutaneous absorption of paraquat from hand, leg, and forearm of humans. *J. Toxicol. Environ. Health, 14*, 759–762.

Wester, R. C., Maibach, H. I., Bucks, D. A. W., Sedik, L., Melendres, J., Liao, C., et al. (1990). Percutaneous absorption of [^{14}C]DDT and benzo[*a*]pyrene from soil. *Fundam. Appl. Toxicol., 15*, 510–516.

Wester, R. C., Maibach, H. I., Melendres, J., Sedik, L., Knaak, J., & Wang, R. (1992a). *In vivo* and *in vitro* percutaneous absorption and skin evaporation of isofenphos in man. *Fundam. Appl. Toxicol., 19*, 521–526.

Wester, R. C., Maibach, H. I., Sedik, L., Melendres, J., Liao, C. L., & DiZio, S. (1992b). Percutaneous absorption of [^{14}C]chlordane from soil. *J. Toxicol. Environ. Health, 35*, 269–277.

Wester, R. C., Maibach, H. I., Sedik, L., Melendres, J., & Wade, M. (1993a). *In vivo* and *in vitro* percutaneous absorption and skin decontamination of arsenic from water and soil. *Fundam. Appl. Toxicol., 20*, 336–340.

Wester, R. C., Maibach, H. I., Sedik, L., Melendres, J., Wade, M., & DiZio, S. (1993b). Percutaneous absorption of pentachlorophenol from soil. *Fundam. Appl. Toxicol., 20*, 68–71.

Wester, R. C., Bucks, D. A., & Maibach, H. I. (1994). Human *in vivo* percutaneous absorption of pyrethrin and piperonyl butoxide. *Food Chem. Toxicol., 32*, 51–53.

Wester, R. C., Melendres, J., Logan, F., Hui, X., Maibach, H. I., Wade, M., et al. (1996a). Percutaneous absorption of 2,4-dichlorophenoxyacetic acid from soil with respect to soil load and skin contact time: *In vivo* absorption in rhesus monkey and *in vitro* absorption in human skin. *J. Toxicol. Environ. Health, 47*, 335–344.

Wester, R. C., Quan, D., & Maibach, H. I. (1996b). *In vitro* percutaneous absorption of model compounds glyphosate and malathion from cotton fabric into and through human skin. *Food Chem. Toxicol, 34*, 731–735.

Wester, R. C., Melendres, J., Sedik, L., Maibach, H. I., & Riviere, J. E. (1998). Percutaneous absorption of salicylic acid, theophylline, 2,4-dimethylamine, diethyl hexyl phthalic acid and *p*-aminobenzoic acid in isolated perfused porcine skin flap compared to man *in vivo*. *Toxicol. Appl. Pharmacol., 1*, 159–165.

Williams, P. L., Thompson, D., Qiao, G. L., Monteiro-Riviere, N. A., Baynes, R. E., & Riviere, J. E. (1996). The use of mechanistically defined chemical mixtures (MDCM) to assess component effects on the percutaneous absorption and cutaneous disposition of topically exposed chemicals. II. Development of a general dermatopharmacokinetic model for use in risk assessment. *Toxicol. Appl. Pharmacol., 141*, 487–496.

Windheuser, J. J., Haslam, J. L., Caldwell, L., & Shaffer, R. D. (1982). The use of *N,N*-diethyl-*m*-toluamide to enhance dermal and transdermal delivery of drugs. *J. Pharm. Sci., 71*, 1211–1213.

Wu, H. X., Evreux-Gros, C., & Descotes, J. (1996). Diazinon toxicokinetics, tissue distribution and anticholinesterase activity in the rat. *Biomed. Environ. Sci., 9*, 359–369.

Zhao, Y. H., Abraham, M. H., Le, J., Hersey, A., Luscombe, C. N., Beck, G., et al. (2002). Rate-limited steps of human oral absorption and QSAR studies. *Pharm. Res., 19*, 1446–1457.

Zhou, S. F. (2008). Structure, function and regulation of P-glycoprotein and its clinical relevance in drug disposition. *Xenobiotica, 38*, 802–832.

CHAPTER 4

Introduction to Biotransformation (Metabolism)

Ernest Hodgson
North Carolina State University, Raleigh, NC, USA

Outline

INTRODUCTION

Williams (1959) first suggested that the metabolism of xenobiotics generally occurs in two phases. The word xenobiotic, however, was coined later, in the mid-1960s, by Dr. Howard Mason to serve as a collective noun including any chemical to which an organism is exposed and which is extrinsic to the normal metabolism of that organism. Thus it includes pesticides, occupational chemicals, clinical drugs, drugs of abuse, deployment-related chemicals, etc., and is a particularly useful term when discussing

Pesticide Biotransformation and Disposition
DOI: 10.1016/B978-0-12-385481-0.00004-6

metabolic pathways and enzymes that have substrates in several of these use classes. Phase I involves predominantly oxidations, reductions, and hydrolysis and serves to introduce a polar group into the molecule. Phase II, consisting primarily of conjugation reactions, involves the combination of the products of phase I reactions with one of several endogenous molecules to form water-soluble, and hence excretable, products.

A number of books review the biotransformation of xenobiotics, either in general or of particular chemical or use classes (e.g., Hodgson, 2010; Jakoby, 1980; Jakoby et al., 1982; Klaassen, 2001; Smart and Hodgson, 2008; Wilkinson, 1976; Williams, 1959). Many treatments of pesticides (e.g., Chambers and Carr, 1995; Ecobichon, 2001; Hodgson and Meyer, 1997, 2010; Hodgson et al., 1995; Kulkarni and Hodgson, 1984a,b; Rose et al., 1999; Smart and Hodgson, 2008) include considerations, not only of pesticide metabolism, but also of the significance of metabolism in the toxicity of pesticides to target and nontarget species.

REACTIONS CATALYZED IN XENOBIOTIC METABOLISM

Many of the chemical reactions involved in the biotransformation of pesticides have now been traced to particular enzymes (see Chapter 5), although some are only inferred from the appearance of derivatives of the parent compound in the tissues or excreta of the dosed animal. Chemical reactions reported to occur in the metabolism of pesticides are summarized in Figure 4.1. It should be noted that biotransformation reactions of pesticides may be either detoxications or activations. Hollingworth et al. (1995) provided an early detailed review of the detection and significance of the active metabolites of pesticides. The biotransformation of most pesticides involves a combination of several chemical reactions and in some instances breakdown products may become part of the general metabolic pool. For example, formaldehyde formed in demethylation reactions may be incorporated into the one-carbon metabolic pool.

XENOBIOTIC-METABOLIZING ENZYMES

There are a large number of both phase I and phase II xenobiotic-metabolizing enzymes, and most exist in the same organism and/or the same tissue as several polymorphic forms (Hodgson, 2010). In vertebrates, the liver is the most important locus for these enzymes although they are found in essentially all tissues.

In the past, most emphasis has been placed on microsomal cytochrome P450 (CYP)-dependent oxidations and reductions of xenobiotics, including pesticides, but other xenobiotic-metabolizing enzymes are found in mitochondria and in the cytosol of hepatocytes and other cells. These are discussed below and in subsequent chapters. More recently, much has been learned of the roles of other phase I enzymes, such as flavin-dependent monooxygenases (FMO), hydrolases, and epoxide hydrolases, and of

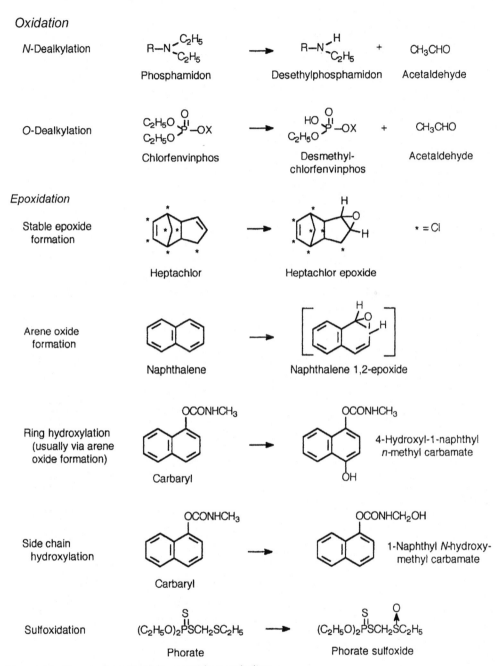

Figure 4.1 Chemical reactions in pesticide metabolism.

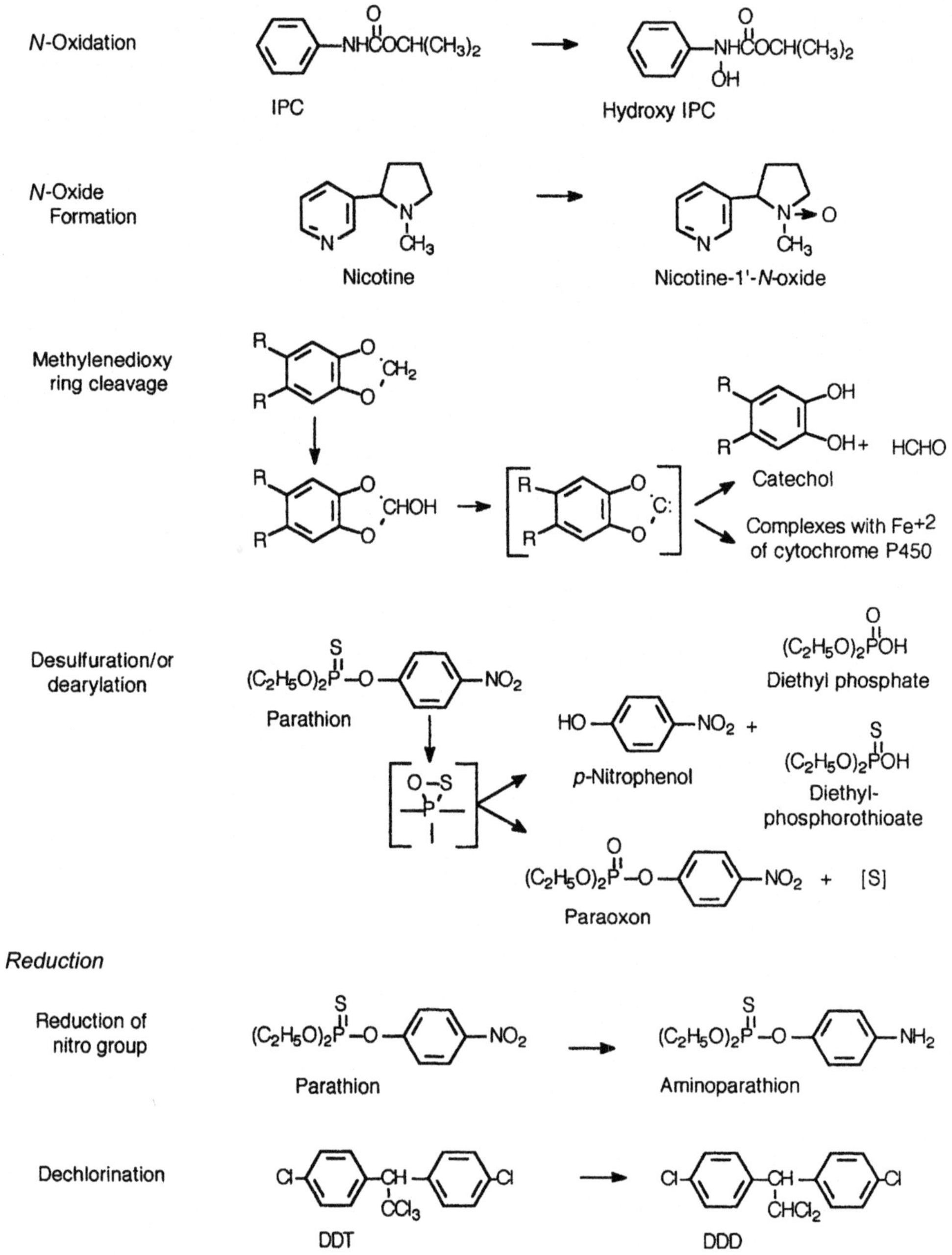

N-Oxidation

IPC   Hydroxy IPC

N-Oxide
Formation

Nicotine   Nicotine-1'-N-oxide

Methylenedioxy
ring cleavage

Catechol

Complexes with Fe+2
of cytochrome P450

Desulfuration/or
dearylation

(C2H5O)2P—O—NO2   (C2H5O)2POH
Parathion   Diethyl phosphate

                                        HO—NO2 +
                                        p-Nitrophenol
                                        (C2H5O)2POH
                                        Diethyl-
                                        phosphorothioate

(C2H5O)2P—O—NO2 + [S]
Paraoxon

Reduction

Reduction of
nitro group

(C2H5O)2P—O—NO2   (C2H5O)2P—O—NH2
Parathion   Aminoparathion

Dechlorination

DDT   DDD

Figure 4.1 (Continued)

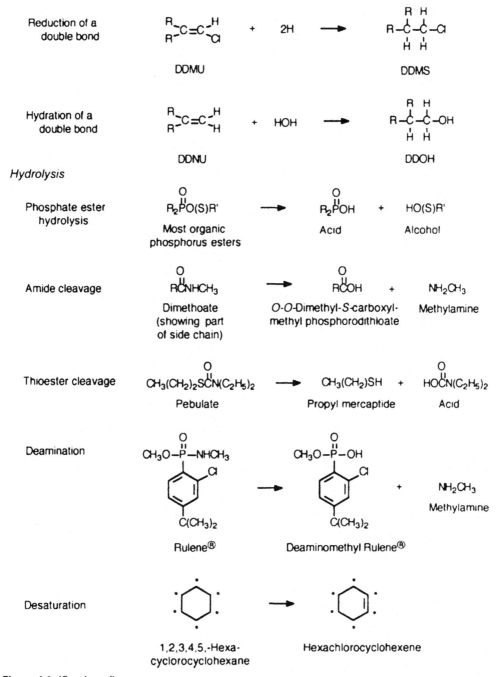

Reduction of a double bond

DDMU   DDMS

Hydration of a double bond

DDNU   DDOH

Hydrolysis

Phosphate ester hydrolysis

Most organic phosphorus esters   Acid   Alcohol

Amide cleavage

Dimethoate (showing part of side chain)   O-O-Dimethyl-S-carboxyl-methyl phosphorodithioate   Methylamine

Thioester cleavage

Pebulate   Propyl mercaptide   Acid

Deamination

Rulene®   Deaminomethyl Rulene®   Methylamine

Desaturation

1,2,3,4,5,-Hexa-cyclorocyclohexane   Hexachlorocyclohexene

Figure 4.1 (Continued)

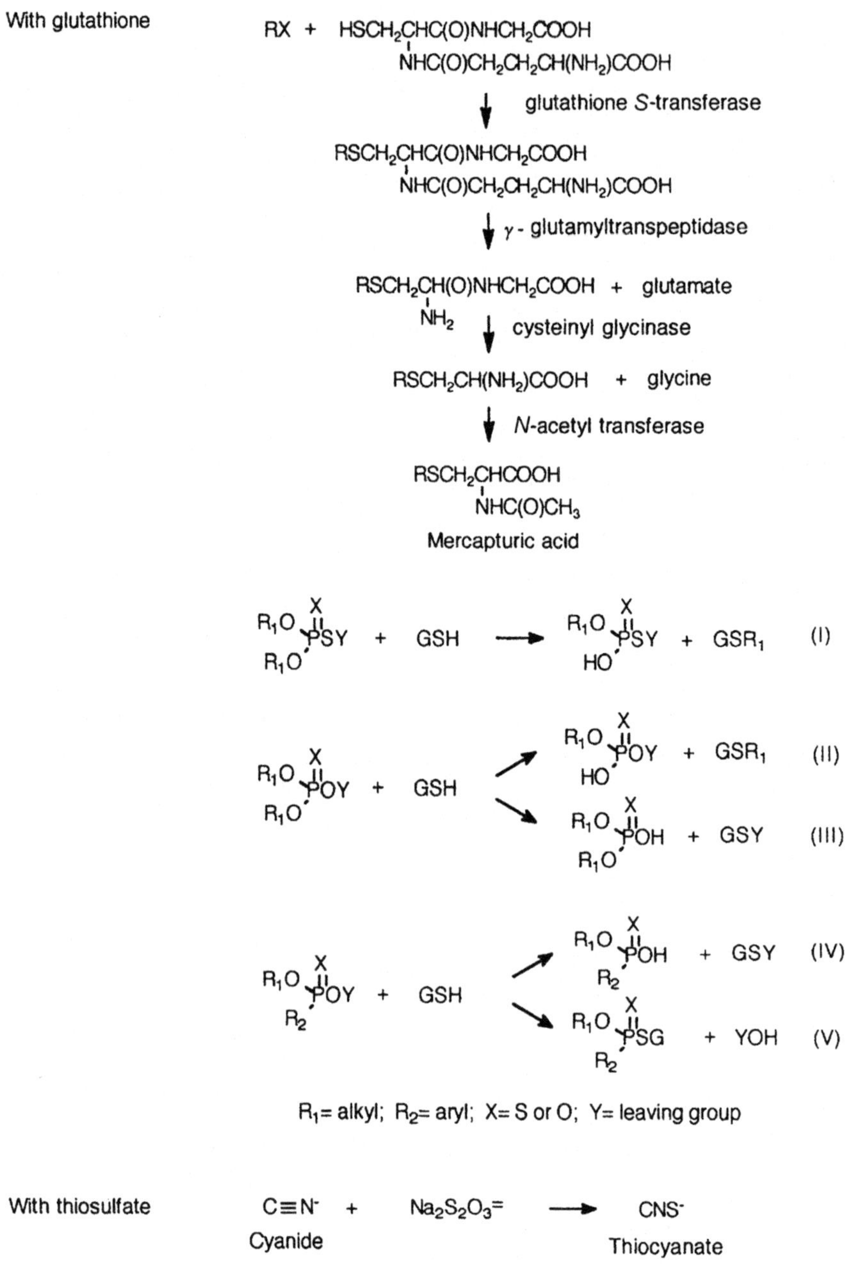

With glutathione

$$RX \ + \ HSCH_2CHC(O)NHCH_2COOH$$
$$NHC(O)CH_2CH_2CH(NH_2)COOH$$

↓ glutathione *S*-transferase

$$RSCH_2CHC(O)NHCH_2COOH$$
$$NHC(O)CH_2CH_2CH(NH_2)COOH$$

↓ γ - glutamyltranspeptidase

$$RSCH_2CH(O)NHCH_2COOH \ + \ glutamate$$
$$NH_2$$

↓ cysteinyl glycinase

$$RSCH_2CH(NH_2)COOH \ + \ glycine$$

↓ *N*-acetyl transferase

$$RSCH_2CHCOOH$$
$$NHC(O)CH_3$$
Mercapturic acid

$$R_1O\overset{X}{\underset{R_1O}{\overset{\|}{P}}}SY \ + \ GSH \ \longrightarrow \ R_1O\overset{X}{\underset{HO}{\overset{\|}{P}}}SY \ + \ GSR_1 \quad (I)$$

$$R_1O\overset{X}{\underset{R_1O}{\overset{\|}{P}}}OY \ + \ GSH$$

$$\nearrow \ R_1O\overset{X}{\underset{HO}{\overset{\|}{P}}}OY \ + \ GSR_1 \quad (II)$$

$$\searrow \ R_1O\overset{X}{\underset{R_1O}{\overset{\|}{P}}}OH \ + \ GSY \quad (III)$$

$$R_1O\overset{X}{\underset{R_2}{\overset{\|}{P}}}OY \ + \ GSH$$

$$\nearrow \ R_1O\overset{X}{\underset{R_2}{\overset{\|}{P}}}OH \ + \ GSY \quad (IV)$$

$$\searrow \ R_1O\overset{X}{\underset{R_2}{\overset{\|}{P}}}SG \ + \ YOH \quad (V)$$

R_1= alkyl; R_2= aryl; X= S or O; Y= leaving group

With thiosulfate

$$C \equiv N^- \ + \ Na_2S_2O_3^= \ \longrightarrow \ CNS^-$$
Cyanide   Thiocyanate

Figure 4.1 (Continued)

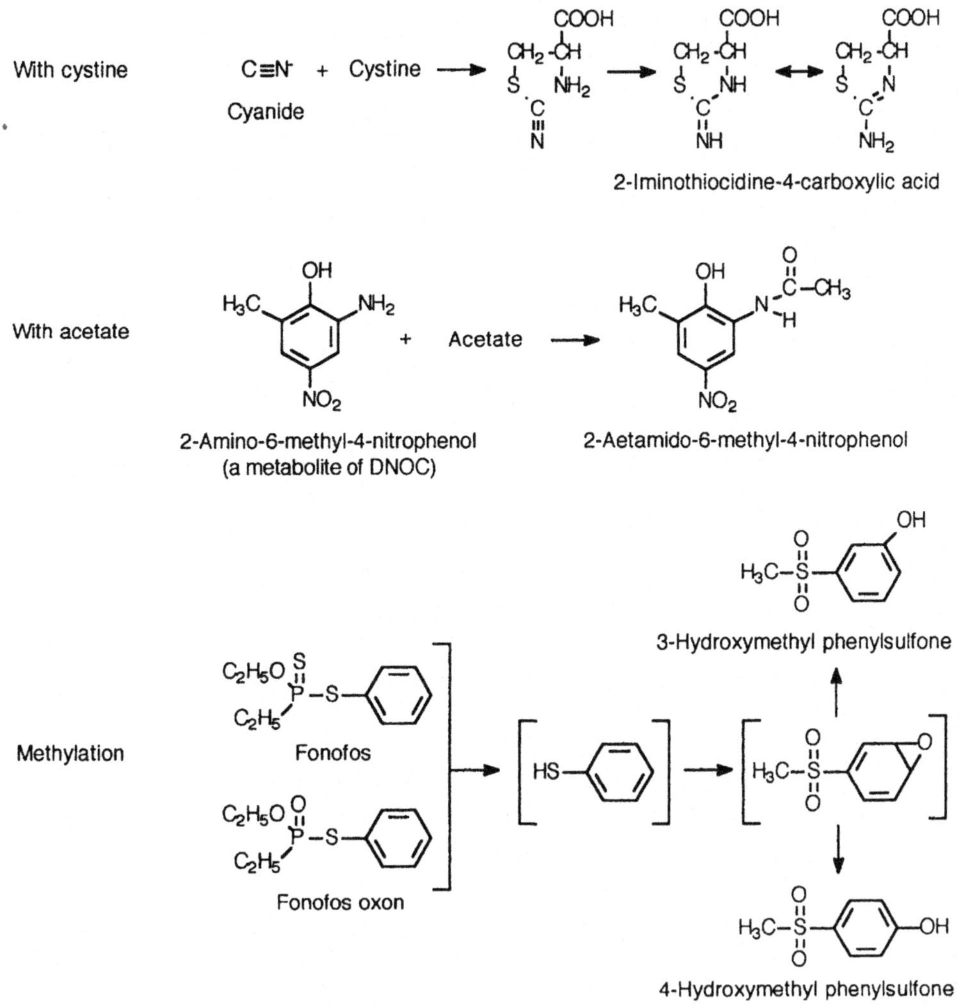

Figure 4.1 (Continued)

cooxidation during prostaglandin synthesis. Emphasis has also been placed on the phase II conjugation reactions as they apply to pesticide metabolism.

By definition, microsomal enzymes are those found in the high-speed particulate microsomal fraction of tissues following homogenization and differential centrifugation. The terms microsomal fraction and microsomes refer to a biochemical preparation and do not correspond to any particular cell structure. However, the major component consists of membranous vesicles derived from the endoplasmic reticulum and its constituent ribosomes. The microsomal fraction consists primarily of rough (with ribosomes) and smooth (without ribosomes) vesicles that correspond to rough and smooth endoplasmic reticulum. The specific activity of smooth microsomes is generally higher than that of rough microsomes for the metabolism of xenobiotics. However, even though rough and smooth microsomes can be separated by density gradient centrifugation, this is generally not done in pesticide metabolism investigations.

The cytosol may be defined as the postmicrosomal supernatant following differential centrifugation of a cell or tissue homogenate.

The majority of studies focusing on pesticide metabolism and the regulation of pesticide-metabolizing enzymes have been conducted in experimental animals, primarily rodents. However, there has been an increase in information about human enzymes, especially the CYP isoforms. Much of this information has been gained through the use of specific substrates, antibodies, human hepatocytes, human cell fractions, and recombinant human enzymes. Studies with human CYPs have become more common and have demonstrated that xenobiotic metabolism and the regulation and expression of xenobiotic-metabolizing enzymes may be quite different in humans and in experimental animals. Such differences make the extrapolation of metabolism studies from experimental animals to humans difficult. It is only as we learn to understand these differences that we can make more accurate and realistic extrapolations to humans.

PHASE I XENOBIOTIC-METABOLIZING ENZYMES

Cytochrome P450 Monooxygenases

Although several enzymes acting in concert may be required for xenobiotic degradation or activation, the initial reaction usually involves a microsomal phase I enzyme catalyzing an oxidation reaction. Reduction reactions, although they may also occur, are relatively uncommon. These enzymes include many of the isoforms of CYP active in the CYP-dependent monooxygenase system, as well as FMO isoforms. The overall aspects of the biochemistry and molecular biology of the CYP system are discussed in detail by Zeldin and Seubert (2008).

Many different pesticide mono-oxygenation reactions are attributed to CYPs, including epoxidation (e.g., aldrin), *N*-dealkylation (e.g., alachlor, atrazine), *O*-dealkylation (e.g., chlorfenvinphos), *S*-oxidation (e.g., phorate), and oxidative desulfuration (e.g., parathion) (Hodgson, 1982–1983; Kulkarni and Hodgson, 1980, 1984a,b). They are discussed in detail in Chapter 5.

Currently the CYP superfamily comprises, in all taxa, over 7000 genes classified into 781 gene families. Of the 110 animal CYP families, 18 are found in vertebrates (Nelson, 2008). The total number of functional CYP genes in any single mammalian species is thought to range from 60 to 200 (Gonzalez, 1990). In vertebrates, most CYP families encode proteins involved primarily in specific endogenous functions (i.e., steroid hormone biosynthesis and metabolism). Other families encode proteins that appear to have more to do with the oxidation of exogenous compounds, such as pesticides, and often display a broad range of substrate specificity (Bogaards et al., 1995; Nebert et al., 1989).

The numerous polymorphisms found in these CYP genes are of considerable importance in the metabolism of xenobiotics, including pesticides. A polymorphism is defined as an inherited monogenetic trait that exists in at least two genotypes (two or more stable alleles) and is stably inherited. They arise as mutational events; if the mutation is in the coding region a variant protein is expressed, thereby affecting the rate and/or extent of metabolism and, potentially, the ratio of different metabolites and the ratio of activation to detoxication.

The details of the interactions of CYPs with xenobiotics have been the subject of intense study for some time, although in these studies clinical drugs have been utilized to a much greater extent than pesticides. Despite the wide range of amino acid sequence homology among CYP isoforms, all these isoforms have a remarkably similar reaction mechanism (Figure 4.2). CYP mono-oxygenation reactions all involve the reduction of one atom of molecular oxygen to water and the incorporation of the other oxygen atom into the substrate. The electrons involved in the reduction of CYP are transferred from NADPH by the NADPH-cytochrome P450 oxidoreductase while, in some cases, the second electron may be derived from NADH via cytochrome b_5.

Reviews of pesticide studies include Hodgson and Kulkarni (1974), Hodgson (1974), and Kulkarni and Hodgson (1980, 1984a,b). Studies of spectral interactions of

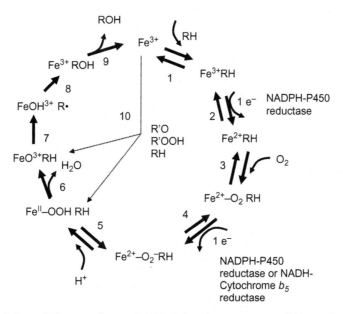

Figure 4.2 Catalytic cycle for cytochrome P450-catalyzed monooxygenation reactions. *(From Zeldin and Seubert, 2008; adapted from F.P. Guengerich, 2001, Chem. Res. Toxicol. 14:611–650.)*

pesticides with CYP, interactions that may be indicative of the ability to act as substrate or inhibitor, have also been carried out (Mailman and Hodgson, 1972; Mailman et al., 1974). Studies using specific isoforms (Hodgson et al., 1998; Levi and Hodgson, 1984, 1988) indicate that even in the same organ of the same species, particular pesticides are metabolized at different rates by different CYP isoforms.

The specificity of various isoforms for pesticide substrates is an area of current interest. Because of the availability of recombinant human isoforms, these studies can now be carried out on human enzymes as well as on those from experimental animals. In an early study of fenitrothion metabolism by mouse liver utilizing four constitutive and two induced CYP isoforms, Levi and Hodgson (1988) showed that all isoforms produced both the cresol detoxication product and the oxon. However, there were significant differences in both overall activity and the oxon/cresol ratio. The most active isoform, induced by phenobarbital and now known as CYP2B10, was also active in the metabolism of parathion and methyl parathion and in all cases produced significantly more oxon than detoxication products. Human CYP3A4 was shown to be most active in the metabolism of parathion, although CYP1A2 and 2B6 also showed activity (Butler and Murray, 1997). In studies on the metabolism of chlorpyrifos by human CYPs (Tang et al., 2001), it has been shown that CYP2B6, CYP2C19, and CYP3A4 are all active, CYP2B6 producing an excess of chlorpyrifos oxon and CYP2C19 an excess of detoxication products, while CYP3A4 produces both in approximately equal quantities.

Studies of the metabolism of triazine herbicides in mice (Adams et al., 1990) and rats (Hanioka et al., 1999) as well as in rats and pigs (Lang et al., 1996) suggested a broad lack of isoform specificity for these substrates. However, Lang et al. (1997) showed that, in humans, CYP1A2 appeared to be the principal, if not the only, isoform responsible for triazine herbicide oxidation. A more recent study (Joo et al., 2010) showed that other CYP isoforms are also be capable of N-dealkylation of atrazine (for details see Chapter 5, Table 5.3). Inui et al. (2000) expressed human CYP1A1, CYP2B6, and CYP2C19 in potatoes and produced resistance to several herbicides, including atrazine, in the host plants, presumably by enabling the plants to metabolize the herbicides.

In studies of chloroacetanilide herbicides (Coleman et al., 1999), it was shown that human CYP3A4 was responsible for the initial O-dealkylation of alachlor. Subsequent studies (Coleman et al., 2000) extended these studies to acetochlor, butachlor, and metachlor. In all cases, CYP3A4 was the most active human isoform, although CYP2B6 also had some activity.

One of the significant features of many of the microsomal CYPs is their inducibility by xenobiotics; thus, stimulation of the metabolism of a chemical by prior administration of the same or another chemical is often taken as presumptive evidence of its metabolism by microsomal enzymes. For example, in mice pretreated with phenobarbital there is an increase in phorate metabolism, suggesting that CYP isoforms, such as

CYP2B or CYP3A, may be important in the metabolism of similar pesticide substrates (Kinsler et al., 1990). Metabolic interactions involving enzyme induction and/or inhibition are discussed in Chapter 7.

CYP-dependent reactions, as they involve pesticides, are considered in detail in Chapter 5. A more mechanism-based classification of CYP-catalyzed xenobiotic oxidations is that of Guengerich and MacDonald (1984). They classified such reactions into six general categories:

1. Carbon hydroxylation: the formation of an alcohol at a methyl, methylene, or methine position.
2. Heteroatom release: the oxidative cleavage of the heteroatom part of a molecule resulting from a hydroxylation adjacent to the heteroatom that generates a geminal hydroxy heteroatom-substituted intermediate such as a carbinolamine, halohydrin, hemiacetal, hemiketal, or hemithioketal. (This intermediate then collapses to release the heteroatom and form a carbonyl compound.)
3. Heteroatom oxygenation: the conversion of a heteroatom-containing substrate to its corresponding heteroatom oxide as in the formation of N-oxides, sulfoxides, or phosphine oxides.
4. Epoxidation: the formation of oxirane derivatives of olefins or aromatic compounds.
5. Oxidative group transfer: a type of reaction that involves a 1,2-carbon shift of a group with the concurrent incorporation of oxygen to form a carbonyl at the C1 position.
6. Olefinic suicide destruction: inactivation of the heme of P450 by an enzyme product.

Flavin-Containing Monooxygenase

The microsomal FMO was known for a number of years as an amine oxidase but was subsequently shown to be also a sulfur oxidase and a phosphorous oxidase. Like CYP, the FMO is a microsomal enzyme, a monooxygenase requiring NADPH and oxygen, and exists as multiple isoforms in various tissues.

However, FMO, unlike CYP, catalyzes only oxygenation reactions, has more specific substrate requirements, and is not known to be subject to induction or inhibition by xenobiotics, apart from competitive inhibition by alternate substrates (Kulkarni and Hodgson, 1984a,b; Ziegler, 1980). The mechanism of catalysis is also distinct (Figure 4.3) in that electrons are transferred directly from NADPH and not via an NADPH-reductase. Also, because the formation of the hydroperoxyflavin form of the enzyme precedes interaction with the substrate, maximum velocity (V_{max}) for a particular FMO isoform is constant for all substrates, although the Michaelis constant (K_m) can vary from one substrate to another. CYP isoforms, on the other hand, show variations from one substrate to another in both V_{max} and K_m. The FMO is found in highest levels in the liver, but is also found in significant levels in the lung and kidney.

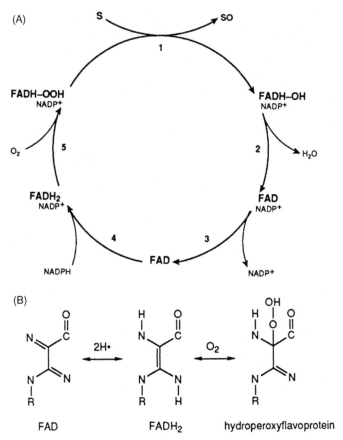

Figure 4.3 The catalytic cycle for flavin-containing monooxygenase-catalyzed monooxygenase reactions.

Recent studies have identified five forms of FMO (FMO1–FMO5), which are differentially expressed with respect to species and tissue (Lawton and Philpot, 1995; Lawton et al., 1994). Each species that has been examined by analysis of genomic DNA appears to contain the same set of FMO genes (Lawton et al., 1994). Although the FMO family possesses multiple isoforms, the number of such forms is small compared to that of the CYP superfamily. While the FMO isoforms all have the same catalytic mechanism, marked differences exist in substrate specificity.

The importance of the FMO in pesticide metabolism was established when it was discovered that the FMO oxidizes a variety of thioether-containing pesticides (Cherrington et al., 1998a,b; Hajjar and Hodgson, 1980, 1982a,b; Levi and Hodgson, 1992; Smyser et al., 1985; Tynes and Hodgson, 1985). It has since been shown that the FMO is capable of oxidative desulfuration (oxon formation) of certain phosphonate insecticides, such as fonofos, through a mechanism distinct from that of oxon formation by CYPs (Smyser and Hodgson, 1985; Smyser et al., 1985) as well as the oxidation of pesticides from a number of different chemical classes (Tynes and Hodgson, 1985). Reviews include Hodgson (1982–1983), Kulkarni and Hodgson (1984a,b), and Hodgson et al. (1998). FMO isoform specificity in pesticide metabolism is also being investigated. For example, Cherrington et al. (1998a,b) showed that, in the mouse, FMO1 metabolizes phorate to phorate sulfoxide but FMO5 is without activity.

A recent review (Hodgson et al., 2008) covers aspects of xenobiotic metabolism by the FMO.

Other Phase I Enzymes
Epoxide Hydrolases
Epoxide rings of certain alkene and arene compounds are hydrated enzymatically by epoxide hydrolases to form the corresponding *trans*-dihydrodiols. The epoxide hydrolases are a family of enzymes known to exist both in the endoplasmic reticulum and in the cytosol. The overall attributes of these enzymes are discussed in Arand et al. (2005) and Hodgson et al. (2008).

Epoxide hydrolases are known to attack xenobiotics of many classes, including some pesticide substrates, although these reactions are subsequent to the initial formation of epoxides. Examples include naphthalene 1,2-oxide and the 3,4- and 5,6-epoxides of carbaryl (Dorough and Casida, 1964) and tridiphane (Magdalou and Hammock, 1987).

Prostaglandin Synthetase
Prostaglandins are synthesized in mammals via a reaction sequence starting with arachidonic acid as substrate. During the second, or peroxidase, step of prostaglandin synthetase action, xenobiotics can be co-oxidized to yield products similar to those formed by various isoforms of CYP (Eling et al., 1983; Hodgson et al., 2008; Marnett and Eling, 1983). A number of pesticides (e.g., aminocarb, parathion) have been shown to act as substrates. These reactions may be important in extrahepatic tissues low in CYP and high in prostaglandin synthetase, such as the seminal vesicle and the inner portion of the medulla of the kidney.

Aldehyde Oxidase
Aldehyde oxidase is a molybdenum-containing oxidoreductase that is similar to xanthine oxidase, also a molybdoenzyme. The extent to which either is involved in pesticide metabolism is uncertain, although it is clear that aldehyde oxidase plays a role in the metabolism of neonicotinoid insecticides (Shi et al., 2009). The metabolism of specific pesticides by this enzyme is summarized in Chapter 5. Aldehyde oxidase is cytosolic and appears, with the exception of the liver, to be expressed in most tissues to a relatively low extent. Aldehyde oxidase oxidizes aldehydes to their corresponding

carboxylic acids, preferring aromatic aldehydes and being relatively inactive toward aliphatic aldehydes such as acetaldehyde. In the presence of a reducing substrate, aldehyde oxidase can catalyze reduction reactions such as azo-reduction and nitro-reduction.

Hydrolases and Amidases

Hydrolase and amidase activities (Hodgson, 2008) are known to be important in the phase I metabolism of xenobiotics, including pesticides. For example, dimethoate is detoxified by amidase activity and the selective toxicity of malathion is due, in large part, to the presence in mammals of carboxylesterases not widely distributed in insects. These enzymes are known from both microsomes and the soluble cytoplasm but are more commonly found in the latter. It appears likely that, in most cases, amidase and esterase activities are different activities of the same enzymes with one or the other activity predominating (Satoh, 1987).

DDT Dehydrochlorinase

In the early 1950s, it was demonstrated that DDT-resistant houseflies detoxified DDT mainly to its noninsecticidal metabolite DDE. The rate of dehydrohalogenation of DDT to DDE was found to vary between various insect strains as well as between individuals. The enzyme involved, DDT dehydrochlorinase, also occurs in mammals but has been studied more intensively in insects.

DDT dehydrochlorinase, a reduced glutathione-dependent enzyme, has been isolated from the 100,000g supernatant of resistant houseflies. Although the enzyme-mediated reaction requires glutathione, the glutathione concentrations are not altered at the end of the reaction and it is a matter of controversy as to whether this enzyme is a cytosolic glutathione transferase.

The lipoprotein enzyme has a molecular mass of 36,000 Da as a monomer and 120,000 Da as the tetramer. The K_m for DDT is 5×10^{27} M with optimum activity at pH 7.4. This enzyme catalyzes the degradation of p,p-DDT to p,p-DDE or the degradation of p,p-DDD (2,2,-bis(p-chlorophenyl)-1,1-dichloroethane) to the corresponding DDT ethylene TDEE (2,2-bis(p-chlorophenyl)-l-chloroethylene). o,p-DDT is not degraded by DDT dehydrochlorinase, suggesting a p,p-orientation requirement for dehalogenation. In general, the DDT resistance of housefly strains is correlated with the activity of DDT dehydrochlorinase, although other resistance mechanisms are known in certain strains.

PHASE II XENOBIOTIC-METABOLIZING ENZYMES

Conjugations may be simple, as in the case of phenol, but often they are more complicated processes in which the final product is derived by several steps. Despite this possible complexity, it is useful to think of the conjugation of xenobiotics as taking place with glucuronic acid to form glucuronides, N-acetylcysteine to form mercapturic acids, glycine to form hippuric and related acids, sulfate to form ethereal sulfates, thiosulfate ions to form thiocyanate, and glutamine to form conjugates of the same name. In fact, the actual conjugations often occur with derivatives of the conjugating molecule, for example, with glutathione, uridine diphosphate glucuronic acid, or phosphoadenine phosphosulfate. Conjugates of foreign chemicals that are rare in mammals, or known only in other classes or phyla, include glucosides, ribosides, ornithines, sulfides, and conjugates with serine, metal complexes, and methylated or acetylated compounds.

Ethereal sulfates, while less important in the metabolism of pesticides than glucuronides, nevertheless may be formed from carbofuran and other carbamates (Dorough, 1968). Glutathione conjugation is important in the metabolism of organophosphates (Motoyama and Dauterman, 1980) and the conjugated products of glutathione adducts may be further metabolized to mercapturic acids, the N-acetylcysteine derivative of the original xenobiotic substrate. Insects and plants are unusual in forming glucosides rather than glucuronides. A general review of phase II metabolism of xenobiotics is that of LeBlanc (2008).

With the exception of glutathione conjugation, most conjugation reactions involving pesticides are secondary, involving, as substrates, the products of phase I reactions. They include glucoside formation, glucuronic acid formation, sulfate formation, and conjugation with amino acids. This area, as it applies to pesticides, is reviewed in Chapter 5.

Glutathione S-Transferases

Conjugation with glutathione, mediated by one of the glutathione S-transferases, is the first step in a sequence leading to a mercapturic acid (Figure 4.1). Several pesticides are metabolized in this way, particularly organophosphorus compounds, DDT, γ-HCH, and organothiocyanates. These reactions and their relationship to pesticides have been reviewed by Motoyama and Dauterman (1980) and by Fukami (1984).

The glutathione S-transferases (GSTs) are an abundant family of dimeric proteins that have the capacity to conjugate glutathione with a variety of compounds containing electrophilic centers. The major hepatic cytosolic GSTs in mammalian liver can be divided into three classes—Alpha (α), Mu (μ), Pi (π)—based on sequence similarity and catalytic activity (Mannervik et al., 1985). Each class may contain one or more functional enzymes. Although all of these classes are capable of binding to a wide variety of pesticides, the Mu class has somewhat higher affinity than the Alpha or Pi class (Dillio et al., 1995; Hayes and Wolf, 1980). Members of the Mu class of GSTs are responsible for conjugating benzo[a]pyrene-7,8-diol-9,10-epoxide as well as a wide variety of pesticides such as the organophosphate insecticides, the halogenated hydrocarbon insecticides, and the S-triazine herbicides (Hayes and Wolf, 1980). Polymorphisms are known to occur in humans in regard to GST enzymes. About 50% of the Caucasian population in the United States is deficient in Mu-class GSTM1.

This polymorphism is due to a deletion in the *GSTM1* gene resulting in the lack of GSTM1 protein formation. Epidemiological studies have implicated this deficiency in an increased risk of lung cancer in smokers, presumably because of the ability of GSTM1 to detoxify chemical carcinogens such as BaP in tobacco smoke (Bell et al., 1992; Nakachi et al., 1993; Seidegard and Pero, 1985; Wormhoudt et al., 1999).

Because conjugation reactions other than those mediated by the GSTs are less well known in the metabolism of pesticides, the enzymatic basis of these conjugations is not discussed in detail. However, this matter has been reviewed by Motoyama and Dauterman (1980), Dorough (1984), Matsumura (1985), and Hollingworth et al. (1995), and several types of conjugation are known to involve pesticides.

Glucuronyl Transferases

Glucuronidation is one of the most important reactions for the elimination of xenobiotics from the body, although not as yet well studied in the case of pesticides or the phase I metabolites of pesticides. Glucuronidation involves the reaction of uridine 5′-diphosphoglucuronic acid with one of a number of possible functional groups, such as R-OH, R-NH$_2$, R-COOH, and others (Hodgson and Rose, 2010).

Glucuronides are, however, important in the metabolism of carbamates such as banol, carbaryl, and carbofuran (Mehendale and Dorough, 1972) as well as some organophosphate compounds (Hutson, 1981) and other chemicals.

Sulfotransferases

Sulfation and sulfate conjugate hydrolysis, catalyzed by various members of the sulfotransferases (SULT) and sulfatase enzyme superfamilies, may play a role in the metabolism and disposition of many xenobiotics. Reactions of the sulfotransferase enzyme with various xenobiotics generally result in the production of water-soluble sulfate esters, which are then eliminated. SULTs, generally speaking, catalyze the sulfation reaction while the sulfatases catalyze the hydrolysis of the sulfate esters formed by the action of the SULTs.

Further details of the biochemical and molecular aspects of SULTs can be found in Hodgson and Rose (2010).

Other Phase II Enzymes
Methyltransferases

There are a number of enzymatic methyltransferase reactions, including N-, O-, and S-methylation, and the substrates for these enzymes may be either xenobiotics or endogenous metabolites. The enzymes involved are briefly described in Hodgson (2010). For almost all methylation reactions the methyl donor is *S*-adenosylmethionine, formed from methionine and ATP.

Cysteine Conjugate β-Lyase

This enzyme uses cysteine conjugates as substrates, releasing the thiol derivative of the original xenobiotic, ammonia, and pyruvic acid, the thiol derivative then undergoing S-methylation to give rise to the methylthio derivative.

Acylation

Acylation reactions are of two types, the first involving transfer of an acetyl group by acetyl-CoA, the second involving activation of the xenobiotic and subsequent reaction with an amino acid. Deacetylation may also occur. These phase II reactions and enzymes have not been widely studied with respect to pesticides.

Phosphate Conjugation

Phosphorylation is not a common reaction in xenobiotic metabolism and, to date, has been described only in insects. The enzyme from the cockroach requires magnesium and utilizes ATP. It is known to phosphorylate 1-naphthol, a metabolite of carbaryl, and *p*-nitrophenol, a metabolite of parathion and methyl parathion.

SUMMARY AND CONCLUSIONS

Pesticides are subject to modification by a wide range of phase I and phase II enzymes and their isoforms and polymorphic variants. The products are numerous and secondary modification of the primary metabolites complicates the situation further. Since the products may be more (activation) or less (detoxication) toxic than the parent chemical, knowledge of metabolism may be critical in the extrapolation from experimental animal to humans necessary for accurate human health risk assessment.

REFERENCES

Adams, N. H., Levi, P. E., & Hodgson, E. (1990). *In vitro* studies of the metabolism of atrazine, simazine and terbutryn in several vertebrate species. *J. Agric. Food Chem., 28*, 1411–1417.

Arand, M., Cronin, A., Adamska, M., & Oesch, F. (2005). Epoxide hydrolases: Structure, function, metabolism and assay. *Methods Enzymol., 400*, 569–588.

Bell, D. A., Thompson, C. L., Taylor, J., Miller, C. R., Perera, F., Hsieh, L. L., et al. (1992). Genetic monitoring of human polymorphic cancer susceptibility genes by polymerase chain reaction: Application to glutathione transferase µ. *Environ. Health Perspect., 98*, 113.

Bogaards, J. J. P., van Ommen, B., Wolf, C. R., & van Bladeren, P. J. (1995). Human cytochrome P450 enzyme selectivities in the oxidation of chlorinated benzenes. *Toxicol. Appl. Pharmacol., 132*, 44–52.

Butler, A. M., & Murray, M. (1997). Biotransformation of parathion in human liver: Participation of CYP3A4 and its inactivation during microsomal parathion oxidation. *J. Pharmacol. Exp. Ther., 280*, 966–973.

Chambers, J. E., & Carr, R. L. (1995). Biochemical mechanisms contributing to species differences in insecticidal toxicity. *Toxicology, 105*, 291–304.

Cherrington, N. J., Can, Y., Cherrington, J. W., Rose, R. L., & Hodgson, E. (1998). Physiological factors affecting protein expression of flavin-containing monooxygenases 1, 3 and 5. *Xenobiotica, 7*, 673–682.

Cherrington, N. J., Falls, J. G., Rose, R. L., Clements, K. M., Philpot, R. M., Levi, P. E., et al. (1998). Molecular cloning, sequence, and expression of mouse flavin-containing monooxygenases 1 and 5 (FMO1 and FMO5). *J. Biochem. Mol. Toxicol., 12*, 205–212.

Coleman, S., Liu, S., Linderman, R., Hodgson, E., & Rose, R. L. (1999). *In vitro* metabolism of alachlor by human liver microsomes and human cytochrome P450 isoforms. *Chem. Biol. Interact., 122*, 27–39.

Coleman, S., Linderman, R., Hodgson, E., & Rose, R. L. (2000). Comparative metabolism of chloracetamide herbicides and selected metabolites in human and rat liver microsomes. *Environ. Health Perspect., 108*, 1151–1157.

Dillio, C., Sacchetta, P., Jannarelli, V., & Aceto, A. (1995). Binding of pesticides to alpha-class, mu-class and pi-class glutathione transferases. *Toxicol. Lett., 76*, 173–177.

Dorough, H. W. (1968). Metabolism of Furadan (NIH-10242) in rats and house-flies. *J. Agric. Food Chem., 16*, 319–325.

Dorough, H. W. (1984). Metabolism of insecticides by conjugation reactions. In F. Matsumura (Ed.), *Differential toxicities of insecticides and halogenated aromatics* (pp. 291–330). New York: Pergamon. (*International encyclopedia of pharmacology and therapeutics*, Sect. 113.)

Dorough, H. W., & Casida, J. E. (1964). Nature of certain carbamate metabolites of the insecticide Sevin. *J. Agric. Food Chem., 12*, 294–304.

Ecobichon, D. J. (2001). Toxic effects of pesticides. In C. D. Klaassen (Ed.), *Casarett and Doull's Toxicology, the basic science of poisons* (6th ed.). New York: McGraw-Hill.

Eling, T., Boyd, J., Reed, G., Mason, R., & Sivarajoh, K. (1983). Xenobiotic metabolism by prostaglandin endoperoxide synthetase. *Drug Metab. Rev., 14*, 1023.

Fukami, J. (1984). Metabolism of several insecticides by glutathione *S*-transferase. In F. Matsumura (Ed.), *Differential toxicities of insecticides and halogenated aromatics* (pp. 223–264). New York: Pergamon. (*International encyclopedia of pharmacology and therapeutics*, Sect. 113.)

Gonzalez, F. J. (1990). Molecular genetics of the P450 superfamily. *Pharmacol. Ther., 45*, 1–38.

Guengerich, F. P., & MacDonald, T. L. (1984). Chemical mechanisms of catalysis by cytochromes P450: A unified view. *Acc. Chem. Res., 17*, 9–16.

Hajjar, N. P., & Hodgson, E. (1980). Flavin adenine dinucleotide-dependent monooxygenase: Its role in the sulfoxidation of pesticides in mammals. *Science, 209*, 1134–1136.

Hajjar, N. P., & Hodgson, E. (1982). Sulfoxidation of thioether-containing pesticides by the flavin-adenine dinucleotide-dependent monooxygenase of pig liver microsomes. *Biochem. Pharmacol., 31*, 745–752.

Hajjar, N. P., & Hodgson, E. (1982). The microsomal FAD-dependent monooxygenase as an activating enzyme: Fonofos metabolism. In R. Synder, D. U. Porke, J. J. Kocsis, D. Jollow, G. G. Gibson, & C. Witmer (Eds.), *Biological reactive intermediates* (Vol. 2, Part B, pp. 1245–1253). New York: Plenum.

Hanioka, N., Jinno, H., Toshiko, T.-K., Nishimura, T., & Ando, M. (1999). In vitro metabolism of chlorotriazines: Characterization of simazine, atrazine, and propazine metabolism using liver microsomes from rats treated with various cytochrome P450 inducers. *Toxicol. Appl. Pharmacol., 156*, 195–205.

Hayes, J. D., & Wolf, C. R. (1980). Role of glutathione in drug resistance. In H. Sies & B. Ketterer (Eds.), *Glutathione conjugation: Its mechanisms and biological significance* (pp. 315–355). London: Academic Press.

Hodgson, E. (1974). Comparative studies of cytochrome P450 and its interaction with pesticides. In M. A. Q. Khan & J. P. Bederka Jr. (Eds.), *Survival in toxic environments* (pp. 213–260). New York: Academic Press.

Hodgson, E. (1982–1983). Production of pesticide metabolites by oxidative reactions. *J. Toxicol. Clin. Toxicol., 19*, 609–621.

Hodgson, E. (2010). *A textbook of modern toxicology* (4th ed.). Hoboken, NJ: John Wiley & Sons.

Hodgson, E., & Kulkarni, A. P. (1974). Interactions of pesticides with cytochrome P450. *ACS Symp. Ser., 2*, 14–38.

Hodgson, E., & Meyer, S. A. (1997). Pesticides. In R. S. McCluskey & D. L. Earnest (Eds.), *Hepatic and gastrointestinal toxicology*. Oxford: Pergamon. (*Comprehensive toxicology*, I. G. Sipes, C. A. McQueen, & A. J. Gandolfi, Series Eds.).

Hodgson, E., & Meyer, S. A. (2010). Pesticides and hepatotoxicity. In R. A. Roth & P. E. Ganey (Eds.), *Hepatic toxicology*. Oxford: Elsevier. (*Comprehensive toxicology*, C. A. McQueen, Series Ed.).

Hodgson, E., & Rose, R. L. (2010). Metabolism of toxicants: *A textbook of modern toxicology* (4th ed.). Hoboken, NJ: John Wiley & Sons.

Hodgson, E., Rose, R. L., Ryu, D.-Y., Falls, G., Blake, B. L., & Levi, P. E. (1995). Pesticide-metabolizing enzymes. *Toxicol. Lett., 82–83*, 73–81.

Hodgson, E., Cherrington, N., Coleman, S. C., Liu, S., Falls, J. G., Cao, Y., et al. (1998). Flavin-containing monooxygenase and cytochrome P450 mediated metabolism of pesticides: From mouse to human. *Rev. Toxicol.*, *2*, 231–243.

Hodgson, E., Das, P. C., Cho, T. M., & Rose, R. L. (2008). Phase I metabolism of toxicants and metabolic interactions. In R. C. Smart & E. Hodgson (Eds.), *Molecular and biochemical toxicology*. Hoboken, NJ: John Wiley & Sons.

Hollingworth, R. M., Kurihara, N., Miyamoto, J., Otto, S., & Paulson, G. D. (1995). Detection and significance of active metabolites of agrochemicals and related xenobiotics in animals. *Pure Appl. Chem.*, *67*, 1487–1532.

Hutson, D. H. (1981). The metabolism of insecticides in man. *Prog. Pestic. Biochem.*, *1*, 287–333.

Inui, H., Kodama, T., Ohkawa, Y., & Ohkawa, H. (2000). Herbicide metabolism and cross-tolerance in transgenic potato plants co-expressing human CYP1A1, CYP2B6, and CYP2C19. *Pestic. Biochem. Physiol.*, *66*, 116–129.

Jakoby, W. B. (Ed.). (1980). *Enzymatic basis of detoxication* (Vols. 1 and 2). New York: Academic Press.

Jakoby, W. B., Bend, J. R., & Caldwell, J. (Eds.). (1982). *Metabolic basis of detoxication*. New York: Academic Press.

Joo, H., Choi, K., & Hodgson, E. (2010). Human metabolism of atrazine. *Pestic. Biochem. Physiol.*, *98*, 73–79.

Kinsler, S., Levi, P. E., & Hodgson, E. (1990). Relative contributions of cytochrome P450 and flavin-containing monooxygenases to the microsomal oxidation of phorate following treatment of mice with phenobarbital, hydrocortisone, acetone, and piperonyl butoxide. *Pestic. Biochem Physiol.*, *37*, 174–181.

Klaassen, C. D. (Ed.). (2001). *Casarett and Doull's toxicology* (6th ed.). New York: McGraw-Hill.

Kulkarni, A. P., & Hodgson, E. (1980). Metabolism of insecticides by the microsomal mixed function oxidase system. *Pharmacol. Ther.*, *8*, 397–475.

Kulkarni, A. P., & Hodgson, E. (1984a). Metabolism of insecticides by the microsomal mixed function oxidase systems. In F. Matsumura (Ed.), *Differential toxicities of insecticides and halogenated aromatics* (pp. 27–128). New York: Pergamon. *(International encyclopedia of pharmacology and therapeutics, Sect. 113.)*

Kulkarni, A. P., & Hodgson, E. (1984b). The metabolism of insecticides: The role of monooxygenase enzymes. *Annu. Rev. Pharmacol. Toxicol.*, *24*, 19–42.

Lang, D., Criegee, D., Grothusen, A., Saalfrank, R. W., & Bocker, R. H. (1996). *In vitro* metabolism of atrazine, terbutylazine, ametryne, and terbutryne in rats, pigs and humans. *Drug Metab. Dispos.*, *24*, 859–865.

Lang, D. H., Rettie, A. E., & Bocker, R. H. (1997). Identification of enzymes involved in the metabolism of atrazine, terbutylazine, ametryne, and terbutryne in human liver microsomes. *Chem. Res. Toxicol.*, *10*, 1037–1044.

Lawton, M. P., & Philpot, R. M. (1995). Emergence of the flavin-containing monooxygenase gene family. *Rev. Biochem. Toxicol.*, *11*, 1–27.

Lawton, M. P., Cashman, J. R., Cresteil, T., Dolphin, E. T., Elfarra, A. A., Hines, R. N., et al. (1994). A nomenclature for the mammalian flavin-containing monooxygenase gene family based on amino acid sequence identities. *Arch. Biochem. Biophys.*, *308*, 254–357.

LeBlanc, G. A. (2008). Phase I—conjugation of toxicants. In R. C. Smart & E. Hodgson (Eds.), *Molecular and biochemical toxicology*. Hoboken, NJ: John Wiley & Sons.

Levi, P. E., & Hodgson, E. (1984). Oxidation of pesticides by purified cytochrome P450 isozymes from mouse liver. *Toxicol. Lett.*, *24*, 221–228.

Levi, P. E., & Hodgson, E. (1988). Stereospecificity of the oxidation of phorate and phorate sulphoxide by purified FAD-containing monooxygenase and cytochrome P450. *Xenobiotica*, *1*, 29–39.

Levi, P. E., & Hodgson, E. (1992). Metabolism of organophosphorus compounds by the flavin-containing monooxygenase. In J. E. Chambers & P. E. Levi (Eds.), *Organophosphates, chemistry, fate, and effects* (pp. 141–154). New York: Academic Press.

Levi, P. E., Hollingworth, R. M., & Hodgson, E. (1988). Differences in oxidative dearylation and desulfuration of fenitrothion by cytochrome P-450 isozymes and in the subsequent inhibition of monooxygenase activity. *Pestic. Biochem. Physiol.*, *32*, 224–231.

Magdalou, J., & Hammock, B. D. (1987). Metabolism of tridiphane (2-(3,5-dichlorophenyl)-2-(2,2,2-trichloroethyl)oxirane) by hepatic epoxide hydrolases and glutathione *S*-transferases in mouse. *Toxicol. Appl. Pharmacol.*, *91*, 439–449.

Mailman, R. B., & Hodgson, E. (1972). The cytochrome P450 substrate optical difference spectra of pesticides with mouse hepatic microsomes. *Bull. Environ. Contam. Toxicol.*, *8*, 186–192.

Mailman, R. B., Kulkarni, A. P., Baker, R. C., & Hodgson, E. (1974). Cytochrome P450 difference spectra: Effect of chemical structure on type II spectra in mouse hepatic microsomes. *Drug Metab. Dispos.*, *2*, 301–311.

Mannervik, B., Alin, P., Guthenberg, C., Jennson, H., Tahir, M. K., Warhom, M., et al. (1985). Identification of 3 classes of cytosolic glutathione transferase common to several mammalian species: Correlation between structural data and enzymatic properties. *Proc. Natl. Acad. Sci. USA*, *82*, 7202.

Marnett, L. J., & Eling, T. E. (1983). Cooxidation during prostaglandin biosynthesis: A pathway for the metabolic activation of xenobiotics. *Rev. Biochem. Toxicol.*, *5*, 135–172.

Matsumura, F. (1985). *Toxicology of insecticides* (2nd ed.). New York: Plenum.

Mehendale, H. M., & Dorough, H. W. (1972). Conjugative metabolism and action of carbamate insecticides. In A. S. Tahori (Ed.), *Insecticide–pesticide chemistry* (pp. 37–49). London: Gordon & Breach.

Motoyama, N., & Dauterman, W. C. (1980). Glutathione *S*-transferases: Their role in the metabolism of organophosphorus insecticides. *Rev. Biochem. Toxicol.*, *2*, 49–70.

Nakachi, K., Inai, K., Hayashi, S., & Kawajiri, K. (1993). Polymorphisms of the CYP1A1 and glutathione *S*-transferase genes associated with susceptibility to lung cancer in relation to cigarette dose in a Japanese population. *Cancer Res.*, *53*, 2994.

Nebert, D. W., Nelson, D. R., & Feyereisen, R. (1989). Evolution of the cytochrome P450 genes. *Xenobiotica*, *19*, 1149–1160.

Nelson, D. R. (2008). <http://drnelson.utmem.edu/p450stats.Feb2008.htm>.

Rose, R. L., Hodgson, E., & Roe, R. M. (1999). Pesticides. In H. Marquardt, S. G. Schafer, R. O. McClellan, & F. Welsch (Eds.), *Toxicology*. San Diego: Academic Press.

Satoh, T. (1987). Role of carboxylases in xenobiotic metabolism. *Rev. Biochem. Toxicol.*, *8*, 155–182.

Seidegard, J., & Pero, R. W. (1985). The hereditary transmission of high glutathione transferase activity toward trans-stilbene oxide in human mononuclear leukocytes. *Hum. Genet.*, *69*, 66.

Shi, X. Y., Dick, R. A., Ford, K. A., & Casida, J. E. (2009). Enzymes and inhibitors in neonicotinoid insecticide metabolism. *J. Agric. Food Chem.*, *57*, 4861–4866.

Smart, R. C., & Hodgson, E. (2008). *Molecular and biochemical toxicology*. Hoboken, NJ: John Wiley & Sons.

Smyser, B. P., & Hodgson, E. (1985). Metabolism of phosphorus-containing compounds by pig liver microsomal FAD-containing monooxygenase. *Biochem. Pharmacol.*, *34*, 1145–1150.

Smyser, B. P., Sabourin, P. J., & Hodgson, E. (1985). Oxidation of pesticides by purified microsomal FAD-containing monooxygenase from mouse and pig liver. *Pestic. Biochem. Physiol.*, *24*, 368–374.

Tang, J., Rose, R. L., Brimfield, A. A., Dai, D., Goldstein, J. A., & Hodgson, E. (2001). Metabolism of chlorpyrifos by human cytochrome P450 isoforms and human, mouse and rat liver microsomes. *Drug Metab. Dispos.*, *29*, 1201–1204.

Tynes, R. E., & Hodgson, E. (1985). Magnitude of involvement of the mammalian flavin-containing monooxygenase in the microsomal oxidation of pesticides. *J. Agric. Food Chem.*, *33*, 471–479.

Wilkinson, C. F. (Ed.). (1976). *Insect biochemistry and physiology*. New York: Plenum.

Williams, R. T. (1959). Detoxication mechanisms. In R. T. Williams (Ed.), *The Metabolism and detoxication of drugs, toxic substances and other organic compounds* (2nd ed.). New York: Wiley.

Wormhoudt, L. W., Cammandeur, J. N. M., & Vermeulen, N. P. E. (1999). Genetic polymorphisms of human *N*-acetyltransferase, cytochrome P450, glutathione-*S*-transferase, and epoxide hydrolase enzymes: Relevance to xenobiotic metabolism and toxicity. *Crit. Rev. Toxicol.*, *29*, 59–124.

Zeldin, D. C., & Seubert, J. M. (2008). Structure, mechanism and regulation of cytochromes P450. In R. C. Smart & E. Hodgson (Eds.), *Molecular and biochemical toxicology*. Hoboken, NJ: John Wiley & Sons.

Ziegler, D. M. (1980). Microsomal flavin-containing monooxygenase: Oxygenation of nucleophilic nitrogen and sulfur compounds. In W. B. Jacoby (Ed.), *Enzymatic basis of detoxication* (Vol. 1). New York: Academic Press (Chap. 9).

CHAPTER 5

Biotransformation (Metabolism) of Pesticides

Ernest Hodgson
North Carolina State University, Raleigh, NC, USA

Outline

INTRODUCTION

The word *metabolism* may be used to designate the sum of chemical reactions that serve to maintain life. Parts of this integrated whole are spoken of as protein metabolism, fat

Pesticide Biotransformation and Disposition
DOI: 10.1016/B978-0-12-385481-0.00005-8

metabolism, nucleic acid metabolism, and the like. Such aspects, as they deal with the processing of the normal endogenous constituents of the body and the effect of pesticides on them, are dealt with in several chapters of the third edition of the *Handbook of Pesticide Toxicology* (Krieger, 2010). The word *metabolism* may also be used to designate the effect of an organism, through its enzymes, on the chemical structure of foreign compounds now more often referred to as xenobiotics. These effects, also called biotransformation, are the subject of this chapter as they apply to pesticides. The enzymes involved in these biotransformations are frequently referred to as xenobiotic-metabolizing enzymes (XMEs).

Given the enormous literature on pesticide metabolism, it is no longer possible to provide an exhaustive review of the subject. The more recent reviews and book chapters referred to throughout are recommended as sources of more detailed and recent information.

It should also be noted that pesticides not only are substrates for XMEs, but may also act as inhibitors or inducers, in either case often with selectivity for specific isoforms. Inhibition and/or induction and interactions consequent to them are considered in Chapter 7.

EXTERNAL TRANSFORMATION

The finding of a derivative of a compound in the tissues or excreta of an animal is not necessarily proof that the compound is the result of biotransformation in that organism. Compounds, especially in thin films, may undergo chemical change when exposed to light or heat. As early as 1961, Mitchell (see Matsumura, 1975) reported the effects of ultraviolet light on 141 pesticides, and Matsumura (1975, 1985) summarized the effects of light and other physical factors on pesticides and their movement in the environment.

The rate and extent of the photochemical degradation of pesticides depend upon the chemical nature of the pesticide, the wavelength of the light, and the presence of other chemicals. These last may act as photosensitizers, forming reactive light-energized intermediates that react with pesticides, or they may react with photoenergized pesticides. The four best known types of photochemical reactions of aromatic pesticides are ring substitution, hydrolysis, oxidation, and polymerization. Examples summarized by Matsumura (1975) include the following: substitution of a ring chlorine in 2,4-D by a hydroxyl group, hydrolysis of carbaryl, oxidation of parathion, and polymerization of pentachlorophenol. A more recent review (Stangroom et al., 2000) discusses the photochemical and thermochemical transformation in water and soil of a number of pesticides, including carbamate, organophosphorus (OP), and pyrethroid insecticides, as well as urea, chlorophenoxy, and triazine herbicides. The reactions involved are often pH-dependent and some may be catalyzed by metal and other ions. A more narrowly focused review (Pehkonen and Zhang, 2002) concerns the degradation of OPs in natural waters.

Table 5.1 (Continued)

Reaction	Example	Reference
	DEF oxidation at C adjacent to S leading to dealkylation	Hur et al. (1992)
Heterocyclic ring hydroxylation	Nicotine→hydroxynicotine	Hucker et al. (1960)
Desulfuration/ dearylation	Parathion→paraoxon	Buratti et al. (2003); Davison (1995); Foxenberg et al. (2007); Kamataki and Neal (1976); Kim et al. (2005)
	Diazinon→diazoxon	Buratti et al. (2003); Poet et al. (2003); Yang et al. (1969, 1971)
	Azinophos-methyl	Buratti et al. (2003)
	Chlorpyrifos	Buratti et al. (2003); Poet et al. (2003); Tang et al. (2001)
	Fenitrothion	Levi et al. (1988)
	Other OPs→oxons	Buratti and Testai (2007); Kulkarni and Hodgson (1984a)
Dehydrogenation	α- and γ-Chlordane→ dichlorochlordene	Chadwick et al. (1975); Street and Blau (1972)
Sulfoxidation	Phorate→phorate sulfoxide→phorate sulfone	Levi and Hodgson (1988)
	Vamidothion→vamidothion sulfoxide	Mehmood M. et al. (1996)
	Thiazopyr→thiazopyr sulfoxide→thiazopyr sulfone	Feng et al. (1994)
	Metam-sodium	Kim et al. (1994); Smyser and Hodgson (1985); Smyser et al. (1985, 1986)
	DEF oxidation of S adjacent to P	Hur et al. (1992)
	Diallate, triallate, and sulfallate	Hackett et al. (1993); Mair and Casida (1991)
	S-oxidation of dithianes, S-oxidation of 7-N,N-dimethylamino-1,2,3,4,5-pentathiocyclooctane	Xia et al. (1995)
	Aldicarb→aldicarb sulfoxide and sulfone	Perkins et al. (1999)
	Methiocarb→methiocarb sulfoxide	Buronfosse et al. (1995)
FMO-DEPENDENT OXIDATIONS		
N-oxidation	Nicotine→nicotine N-oxide	Tynes and Hodgson (1985a,b)
	Tetram→tetram N-oxide	Hajjar and Hodgson (1980, 1982a)
Sulfoxidation	Phorate→phorate sulfoxide	Hajjar and Hodgson (1982b); Kim et al. (1994); Levi and Hodgson (1988)
	Methiocarb→methiocarb sulfoxide	Buronfosse et al. (1995); Furnes and Schlenk (2005)
	Demeton-O	Furnes and Schlenk (2005)
	Ethiofencarb	Furnes and Schlenk (2005)

(Continued)

Table 5.1 (Continued)

Reaction	Example	Reference
Oxidative desulfuration	Fonofos→fonofos oxon	Furnes and Schlenk (2005); Hajjar and Hodgson (1980, 1982a); Smyser and Hodgson (1985); Smyser et al. (1985, 1986)
Nitro reduction	Parathion→aminoparathion	Hitchcock and Murphy (1967)
	Neonicotinoids	Schultz-Jander and Casida (2002); Shi et al. (2009)
DECHLORINATION		
	DDT→TDE	Esaac and Matsumura (1984)
HYDROLYSIS		
	DDVP→desmethyl DDVP	Hodgson and Casida (1962)
	Deltamethrin→3-2,2-dibromovinyl-2,2-cyclopropane carboxylic acid and 3-phenoxybenzalde	Akhtar (1984); Anand (2006); Godin et al. (2006, 2007); Ross et al. (2006); Ross and Crow (2007)
	Bioresmethrin	Ross et al. (2006)
	Esfenvalerate	Godin et al. (2007)
	Permethrin	Crow et al. (2007); Ghiasuddin and Soderlund (1984)
	Pyrethroids ND pyrethroid-like model substrates	Huang et al. (2005); Stok et al. (2004)
	OP hydrolase hydrolysis of P–S bond in acephate, azinphosmethyl, demeton-S, malathion, and phosalone	Lai et al. (1995)
	Malathion→desethyl malathion	Buratti and Testai (2005)
CONJUGATION		
Glucuronidation	Dieldrin→dieldrin glucuronide	Baldwin et al. (1972); Hutson (1976); Matthews and Matsumura (1969)
	Methoxychlor	Hazai et al. (2004)
	Carbaryl→naphthyl glucuronide	Chin et al. (1979a,b,c)
	Neonicotinoid metabolites	Shi et al. (2009)
Sulfation	Carbaryl→naphthyl sulfate	Chin et al. (1979a,b,c)
Acetylation	Fluoroacetamide→fluoroacetyl CoA	Peters (1963)
Glutathione conjugation	Methyl parathion→GSH desmethyl methyl parathion-1-methyl GSH	Abel et al. (2004a); Choi et al. (2006); Hollingworth (1969)
Atrazine	GSH conjugates	Abel et al. (2004b)
Methylation	Nicotinoid metabolites	Shi et al. (2009)
EPOXIDE HYDROLASE		
	Tridiphane	Magdalou and Hammock (1987)

associations between pesticides and susceptibility to various pesticide health effects have provided information about the functional consequences of metabolism by affected enzymes. A summary of some of the methods used to demonstrate XMEs in in vitro studies is shown in Table 5.2.

Biotransformation in the Liver

Pesticide absorption occurs through the skin as well as the respiratory and gastrointestinal tracts, with eventual disposition to the liver from all routes of exposure, the liver being the primary site of pesticide biotransformation for the purpose of facilitating clearance through excretion of water-soluble products and detoxication. However, the

Table 5.2 In Vitro Tests for Microsomal Xenobiotic-Metabolizing Enzymes

Reaction or enzyme	Substrate	Reference
Ester hydrolysis	Methylthiobutyrate, p-nitrophenyl acetate	Heymann and Mentlein (1981)
CYTOCHROME P450		
1A1 rat	Benzoapyrene	Gelboin and Conney (1968); Denison et al. (1983)
	Ethoxyresorufin	Aitio (1978); Burke and Mayer (1974); Pohl and Fouts (1980)
1A2 rat	Acetanilide	Lewandowski et al. (1990); Mitoma and Udenfriend (1962)
1A2 human	Phenacetin	Xenotech (2008)
2A6	Coumarin	Xenotech (2008)
2B1/2 rat	Pentoxyresorufin	Lubet et al. (1985, 1990)
	Benzphetamine	Werringloer (1978)
	Testosterone, 16α,16β-hydroxylation	Sonderfan et al. (1987); Wood et al. (1983)
2B6 human	Buproprion	Xenotech (2008)
2C8 human	Amodiaquine	Xenotech (2008)
2C9 human	Diclofenac	Xenotech (2008)
2C19 human		Xenotech (2008)
2D6 human	Dextromethorphan	Xenotech (2008)
2E1 rat	p-Nitrophenol	Koop (1986)
2E1	Chloroxazone	Xenotech (2008)
3A1 rat	Testosterone, 6β-hydroxylation	Li et al. (1995); Sonderfan et al. (1987)
3A4/5 human	Testosterone 6,β-hydroxylation	Xenotech (2008)
4A1 rat, human	Lauric acid	Kinsler et al. (1988); Xenotech (2008)
FLAVIN-CONTAINING MONOOXYGENASE		
	Phorate	Levi and Hodgson (1988)
	N,N-dimethylaniline	Tynes and Hodgson (1985a,b)
	Methimazole	Dixit and Roche (1984)
	Thiobenzamide	Cashman and Hanzlik (1981)

high level of oxidative metabolism in the liver makes this organ a possible target for more toxic metabolic product activation when detoxifying and protective mechanisms are overwhelmed. Both acute pesticide poisoning with liver involvement and chronic liver toxicity, including cancer, have been associated with pesticide exposure.

CYP and FMO Monooxygenation

As illustrated in Table 5.1, CYP carries out many different types of monooxygenation of pesticide substrates, such as epoxidation (e.g., aldrin), N-dealkylation (e.g., atrazine), O-dealkylation (e.g., chlorfenvinphos), sulfoxidation (e.g., phorate), and oxidative desulfuration (e.g., parathion) (Ecobichon, 2001; Kulkarni and Hodgson, 1980, 1984a,b; Hodgson and Meyer, 2010). Substrates for FMO are similarly diverse, but all are soft nucleophiles, a category that includes many organic chemicals containing sulfur, nitrogen, phosphorus, or selenium heteroatoms. Although CYP isoforms appear to prefer hard nucleophiles as substrates, there is considerable overlap, and most, if not all, substrates for FMO are also CYP substrates. The reverse, however, is not true, since oxidation at carbon atoms is readily catalyzed by CYP but rarely, if at all, by FMO. Moreover, even when the same substrate is oxidized by both CYP and FMO, there may be differences in the rate of oxidation, in the products, or in the stereochemistry of the same product. While isoforms of both CYP and FMO are expressed in the liver, they are broadly expressed in other organs, the proportions of different isoforms varying from organ to organ. Pesticide substrates for FMO include organophosphates such as phorate, disulfoton, and demeton-O, which yield sulfoxides; the phosphonate fonofos, which yields fonofos oxon; carbamates such as aldicarb, methiocarb, and ethiofencarb; dithiocarbamate herbicides such as sodium metham; botanical insecticides such as nicotine; and cotton defoliants such as the trivalent organophosphorus defoliant folex (Buronfosse et al., 1995; Furnes and Schlenk, 2005; Krueger and Williams, 2005; Smyser and Hodgson, 1985, 1986; Smyser et al., 1985; Tynes and Hodgson, 1985a,b; Venkatesh et al., 1992a).

Over 7500 animal CYP isoforms in 781 gene families have been characterized across all taxa, and genomic and protein sequences are known. A system of nomenclature based upon derived amino acid sequences was proposed in 1987, and entries are continuously updated (http://drnelson.utmem.edu/CytochromeP450.html). The degree of similarity in sequence classifies members into a CYP numeric gene family and then a letter subfamily (Nelson, 2006), such that individual isoforms have unique CYP number-letter-number annotations, e.g., CYP1A1. Of the 110 animal CYP families, 18 are found in vertebrates (Nelson, 2006). The total number of functional CYP genes in any single mammalian species is thought to range from 60 to 200 (Gonzalez, 1990). Whereas some CYP isoforms are substrate specific, those involved in xenobiotic metabolism tend to be relatively nonspecific, although substrate preferences are usually evident. FMOs, like CYPs, are located in the endoplasmic reticulum of hepatocytes and other vertebrate cells and catalyze the NADPH-dependent monooxygenation of pesticides, especially those with N, S, or P heteroatoms (Cashman and Zhang, 2006; Ziegler, 2002).

Early studies on the contributions of individual CYP isoforms using partially purified CYP preparations from mouse liver showed considerable variation between fractions in oxidation of pesticide substrates, in spectral binding, and in inhibition by piperonyl butoxide (Beumel et al., 1985; Levi and Hodgson, 1985). Subsequently, the use of highly purified CYPs from the livers of phenobarbital and β-naphthoflavone-treated mice showed that these fractions had much higher activity toward the organophosphorus insecticides fenitrothion, parathion, and methyl parathion than did similar fractions from the livers of untreated mice, suggesting the importance of the CYP1A and CYP2B families in these oxidations. The isoforms also produced different amounts of detoxication products compared with the more toxic oxons, with CYP2Bs forming more of the oxon (Levi et al., 1988). Similar studies showed the importance of the CYP2B family in the hepatic metabolism of phorate to phorate sulfoxide (Kinsler et al., 1988, 1990). More recently, it has become clear that CYP2B6 is one of the most important CYP isoforms in the human metabolism of pesticides (Croom et al., 2009, 2010; Hodgson and Rose, 2007; Tang et al., 2001). This is despite the fact that CYP3A4 is always the most abundant CYP isoform in the human liver, as the kinetic constants for CYP2B6 favor its greater involvement.

Studies of in vitro metabolism of pesticides were, until recently, carried out only on surrogate animals. During the past decade, however, because of the availability of human liver cells, cell fractions, and recombinant human XMEs, there has been an increasing number of studies of the human metabolism of pesticides, and in some instances, variations due to polymorphisms have been demonstrated. A summary of pesticide substrates for human hepatic XMEs is presented in Table 5.3. It is apparent that essentially all of the human xenobiotic-metabolizing CYPs, as well as some other phase I enzymes, have one or more pesticide substrates. A number of studies have shown the importance of both the relative amounts of different CYP or FMO isoforms present (Buratti and Testai, 2005, 2007; Buratti et al., 2002, 2003, 2007; Cashman and Zhang, 2006; Cherrington et al., 1998b; Mutch et al., 1999, 2003; Yang et al., 2009; Tang et al., 2001, 2002, 2004; Usmani et al., 2002, 2004a,b; Scollon et al., 2009) and the effects of polymorphisms on the extent of metabolism and the distribution of metabolites (Dai et al., 2001; Tang et al., 2001). Additional references can be found in Hodgson (2003).

Other Phase I Enzymes

Epoxide hydrolase is another phase I enzyme known to metabolize pesticides, a well-known example being the metabolism of the herbicide tridiphane by the epoxide hydrolase of mouse liver (Magdalou and Hammock, 1987).

The role of carboxylesterases, CYPs, and alcohol and aldehyde dehydrogenases in the hepatic metabolism of pyrethroids has recently been studied in both rodents

Table 5.3 Human Phase I Xenobiotic-Metabolizing Enzymes Active in Pesticide Metabolism

Phase I isoform	Substrate	Reference
Alcohol dehydrogenase-α	*Permethrin metabolite:* phenoxybenzyl alcohol	Choi et al. (2002)
Alcohol dehydrogenase β-I	*Permethrin metabolite:* phenoxybenzyl alcohol	Choi et al. (2002)
Alcohol dehydrogenase β-II	*Permethrin metabolite:* phenoxybenzyl alcohol	Choi et al. (2002)
Alcohol dehydrogenase-γ	*Permethrin metabolite:* phenoxybenzyl alcohol	Choi et al. (2002)
Aldehyde dehydrogenase		
Aldehyde oxidase	Neonicotinoids	Dick et al. (2005); Honda et al. (2006)
ALDH3A1	*Permethrin metabolite:* phenoxybenzyl aldehyde	Choi et al. (2002)
CYP1A1	*Insecticides*	
	Carbaryl	Tang et al. (2002)
	Carbofuran	Usmani et al. (2004a)
	Sulprofos	Usmani et al. (2004b)
	Herbicides	
	Diuron	Abass et al. (2007a)
	Ametryne	Lang et al. (1996, 1997)
	Atrazine	Joo et al. (2010); Lang et al. (1996, 1997)
	Terbuthylazine	Lang et al. (1996, 1997)
	Terbutryne	Lang et al. (1996, 1997)
	Insect repellent: DEET	Usmani et al. (2002)
CYP1A2	*Anthelminthic:* thiobendazole	Coulet, 1998
	Insecticides	
	Azinphos-methyl	Buratti et al. (2002, 2003)
	Carbaryl	Tang et al. (2002)
	Carbofuran	Usmani et al. (2004a)
	Chlorpyrifos	Buratti et al. (2002, 2003); Foxenberg et al. (2007); Tang et al. (2001); Mutch and Williams (2006); Sams et al. (2004)
	Diazinon	Buratti et al. (2002, 2003); Mutch and Williams (2006)
	Disulfoton	Usmani et al. (2004b)
	Fenthion	Leoni et al. (2008)
	Imidacloprid	Schultz-Jander and Casida (2002)
	Malathion	Buratti et al. (2005)
	Methiocarb	Usmani et al. (2004b)
	Parathion	Buratti et al. (2003); Butler and Murray (1997); Foxenberg et al. (2007); Mutch et al. (1999, 2003, 2006); Sams et al. (2000)
	Phorate	Hodgson et al. (1998); Usmani et al. (2004b)
	Sulprofos	Usmani et al. (2004b)

(Continued)

Table 5.3 (Continued)

Phase I isoform	Substrate	Reference
	Methoxychlor	Hu and Kupfer (2002a,b)
	Herbicides	
	Ametryne	Lang et al. (1996, 1997)
	Atrazine	Lang et al. (1996, 1997); Joo et al. (2010)
	Terbuthylazine	Lang et al. (1996, 1997)
	Terbutryne	Lang et al. (1996, 1997)
	Diuron	Abass et al. (2007a)
CYP1B1	Carbofuran	Usmani et al. (2004a,b)
CYP2A6	*Insecticides*	
	Carbaryl	Tang et al. (2002)
	Dimethoate	Buratti and Testai (2007)
	Imidachloprid	Schultz-Jander and Casida (2002)
	Methoxychlor	Hu and Kupfer (2002a,b)
	Nicotine .	Tyndale and Sellers (2002)
	Insect repellent:	Usmani et al. (2002)
	DEET *Fungicide*	Abass et al. (2007c)
	Metaloxyl	
CYP2B6	*Insecticides*	
	Azinophos-methyl	Buratti et al. (2002, 2003)
	Carbaryl	Tang et al. (2002)
	Carbofuran	Usmani et al. (2004a,b)
	Carbosulfan	Abass et al. (2009, 2010)
	Chlorpyrifos	Buratti et al. (2002, 2003); Sams et al. (2004); Foxenberg et al. (2007); Mutch and Williams (2006); Tang et al. (2001)
	Diazinon	Buratti et al. (2002, 2003); Sams et al. (2004); Mutch and Williams (2006)
	Dimethoate	Buratti et al. (2007)
	Disulfoton	Usmani et al. (2004b)
	Endosulfan	Casabar et al. (2006); Lee et al. (2006, 2007)
	Imidacloprid	Schultz-Jander and Casida (2002)
	Methiocarb	Usmani et al. (2004b)
	Parathion	Buratti et al. (2003); Butler and Murray (1997); Foxenberg et al. (2007); Mutch et al. (1999, 2003, 2006); Sams et al. (2000)
	Phorate	Usmani et al. (2004b)
	Profenofos	Abass et al. (2007b)
	Insect repellent: DEET	Usmani et al. (2002)
	Herbicides	
	Acetachlor	Coleman et al. (2000)
	Alachlor	Coleman et al. (2000)
	Ametryne	Lang et al. (1996, 1997)
	Atrazine	Lang et al. (1996, 1997)
	Butachlor	Lang et al. (1996, 1997)

(*Continued*)

Table 5.3 (Continued)

Phase I isoform	Substrate	Reference
	Metolachlor	Lang et al. (1996, 1997)
	Terbutryne	Lang et al. (1996, 1997)
CYP2C8	*Insecticides*	
	Carbaryl	Tang et al. (2002)
	Carbofuran	Usmani et al. (2004a)
	Chlorpyrifos	Mutch and Williams (2006)
	Dimethoate	Buratti et al. (2007)
	Parathion	Mutch et al. (2003, 2006)
	Phorate	Hodgson et al. (1998)
	Deltamethrin, esfenvalerate	Godin et al. (2006, 2007)
	Methoxychlor	Hu and Kupfer (2002a,b)
	Herbicides	
	Ametryne	Lang et al. (1997)
	Atrazine	Joo et al. (2010)
CYP2C9	*Insecticides*	
	Chlorpyrifos	Foxenberg et al. (2007); Tang et al. (2001)
	Dimethoate	Buratti et al. (2007)
	Esfenvalerate	Godin et al. (2007)
	Fenthion	Leoni et al. (2008)
	Imidacloprid	Schultz-Jander and Casida (2002)
	Malathion	Buratti et al. (2005)
	Methoxychlor	Bikadi and Hazai (2008)
	Parathion	Foxenberg et al. (2007)
	Phorate	Hodgson et al. (1998)
	Herbicide: ametryne	Lang et al. (1997)
CYP2C9*1	*Insecticides*	
	Carbaryl	Tang et al. (2002)
	Disulfoton	Usmani et al. (2004b)
	Endosulfan-α	Casabar et al. (2006)
	Methiocarb	Usmani et al. (2004b)
	Phorate	Usmani et al. (2004b)
	Sulprofos	Usmani et al. (2004b)
	Methoxychlor	Hu and Kupfer (2002)
CYP2C9*2	*Insecticides*	
	Carbaryl	Tang et al. (2002)
	Disulfoton	Usmani et al. (2004b)
	Sulprofos	Usmani et al. (2004b)
CYP2C9*3	*Insecticides*	
	Carbaryl	Tang et al. (2002)
	Carbofuran	Usmani et al. (2004a,b)
	Sulprofos	Usmani et al. (2004b)
CYP2C18	*Insecticides*	
	Carbaryl	Tang et al. (2002)
	Disulfoton	Usmani et al. (2004b)

(*Continued*)

Table 5.3 (Continued)

Phase I isoform	Substrate	Reference
	Phorate	Usmani et al. (2004b)
	Sulprofos	Usmani et al. (2004b)
CYP2C19	*Insecticides*	
	Azinphos-methyl	Buratti et al. (2002)
	Carbaryl	Tang et al. (2002)
	Carbofuran	Usmani et al. (2004a)
	Chlorpyrifos	Buratti et al. (2002); Sams et al. (2004); Mutch and Williams (2006); Foxenberg et al. (2007); Tang et al. (2001)
	Deltamethrin	Godin et al. (2007)
	Esfenvalerate	Godin et al. (2007)
	Diazinon	Buratti et al. (2002); Kappers et al. (2001); Mutch and Williams (2006); Sams et al. (2004)
	Dimethoate	Buratti et al. (2007)
	Disulfoton	Usmani et al. (2004b)
	Endosulfan-α	Casabar et al. (2006)
	Fenthion	Leoni et al. (2008)
	Fipronil	Tang et al. (2004)
	Imidacloprid	Schultz-Jander and Casida (2002)
	Malathion	Buratti et al. (2005)
	Methiocarb	Usmani et al. (2004b)
	Parathion	Buratti et al. (2002); Foxenberg et al. (2007); Mutch et al. (2003, 2006)
	Phorate	Hodgson et al. (1998); Usmani et al. (2004a,b)
	Profenofos	Abass et al. (2007b)
	Sulprofos	Usmani et al. (2004b)
	Deltamethrin	Godin et al. (2007)
	Esfenvalerate	Godin et al. (2007)
	Methoxychlor	Bikadi and Hazai (2008); Hu and Kupfer (2002)
	Insect repellent: DEET	Usmani et al. (2002)
	Herbicides	
	Ametryne	Lang et al. (1996, 1997)
	Atrazine	Lang et al. (1996, 1997)
	Terbuthylazine	Lang et al. (1996, 1997)
	Diuron	Abass et al. (2007a)
CYP2C19*1B	*Insecticide:* chlorpyrifos	Tang et al. (2001)
CYP2C19*8	*Insecticide:* chlorpyrifos	Tang et al. (2001)
CYP2C19*6	*Insecticide:* chlorpyrifos	Tang et al. (2001)
CYP2C19*5	*Insecticide:* chlorpyrifos	Tang et al. (2001)
CYP2D*1	*Insecticides*	
	Carbaryl	Tang et al. (2002)
	Carbofuran	Usmani et al. (2004b)
	Disulfoton	Usmani et al. (2004b)

(Continued)

Table 5.3 (Continued)

Phase I isoform	Substrate	Reference
	Sulprofos	Usmani et al. (2004b)
	Methoxychlor	Hu and Kupfer (2002)
	Insect repellent: DEET	Usmani et al. (2002)
CYP2D6	*Insecticides*	
	Diazinon	Sams et al. (2000)
	Imidacloprid	Schultz-Jander and Casida (2002)
	Methiocarb	Usmani et al. (2004b)
	Parathion	Mutch et al. (2003)
	Herbicide: atrazine	Lang et al. (1997)
CYP2E1	*Insecticides*	
	Carbaryl	Tang et al. (2002)
	Imidacloprid	Schultz-Jander and Casida (2002)
	Parathion	Mutch et al. (2003, 2006)
	Phorate	Hodgson et al. (1998); Usmani et al. (2004b)
	Insect repellent: DEET	Usmani et al. (2002)
	Herbicide: atrazine	Lang et al. (1997)
CYP3A4	*Fungicide:* metalaxyl	Abass et al. (2007c)
	Insecticides	
	Azinphos-methyl	Buratti et al. (2002, 2003)
	Carbaryl	Tang et al. (2002)
	Carbofuran	Usmani et al. (2004a)
	Carbosulfan	Abass et al. (2010)
	Chlorpyrifos	Buratti et al. (2002, 2003, 2006); Dai et al. (2001); Sams et al. (2004); Mutch and Williams (2006); Foxenberg et al. (2007); Tang et al. (2001)
	Deltamethrin	Godin et al. (2007)
	Diazinon	Buratti et al. (2003); Kappers et al. (2001); Sams et al. (2000); Mutch and Williams (2006)
	Dimethoate	Buratti and Testai (2007)
	Disulfoton	Usmani et al. (2004b)
	Endosulfan-α	Casabar et al. (2006); Lee et al. (2006, 2007)
	Endosulfan-β	Lee et al. (2006, 2007)
	Esfenvalerate	Godin et al. (2007)
	Fenthion	Buratti et al. (2006); Leoni et al. (2008)
	Fenthion sulfoxide	Buratti et al. (2006)
	Fipronil	Tang et al. (2004)
	Imidacloprid	Schultz-Jander and Casida (2002)
	Malathion	Buratti et al. (2005, 2006)
	Methiocarb	Usmani et al. (2004b)
	Parathion	Buratti et al. (2003, 2006); Butler and Murray (1997); Foxenberg et al. (2007); Kappers et al. (2001); Mutch et al. (1999, 2003, 2006); Sams et al. (2000)
	Phorate	Hodgson et al. (1998); Usmani et al. (2004b)

(Continued)

Table 5.3 (Continued)

Phase I isoform	Substrate	Reference
	Profenofos	Abass et al. (2007a,b,c)
	Sulprofos	Usmani et al. (2004b)
	Thiomethoxam	Honda et al. (2006)
	Trichlorfon	Fujioka and Casida (2007)
	Vamidothion	Mehmood, Z. et al. (1996)
	Methoxychlor	Hu and Kupfer (2002)
	Insect repellent: DEET	Usmani et al. (2002)
	Herbicides	
	Acetachlor	Coleman et al. (1999, 2000); Hodgson et al. (1998)
	Alachlor	Coleman et al. (1999, 2000); Hodgson et al. (1998)
	Ametryne	Lang et al. (1996, 1997)
	Atrazine	Lang et al. (1996, 1997); Joo et al. (2010)
	Butachlor	Lang et al. (1996, 1997)
	Terbuthylazine	Lang et al. (1996, 1997)
	Terbutryne	Lang et al. (1996, 1997)
	Diuron	Abass et al. (2007a)
	2,4-D	Mehmood et al. (1996a,b)
CYP3A4–F189S	*Insecticide:* chlorpyrifos	Dai et al. (2001)
CYP3A4–L293P	*Insecticide:* chlorpyrifos	Dai et al. (2001)
CYP3A4–M445T	*Insecticide:* chlorpyrifos	Dai et al. (2001)
CYP3A4–P467S	*Insecticide:* chlorpyrifos	Dai et al. (2001)
CYP3A5	*Insecticides*	
	Carbaryl	Tang et al. (2002)
	Carbofuran	Usmani et al. (2004a)
	Chlorpyrifos	Foxenberg et al. (2007)
	Deltamethrin	Godin et al. (2007)
	Diazinon	Mutch and Williams (2006)
	Endosulfan-β	Lee et al. (2006, 2007)
	Fenthion sulfoxide	Buratti et al. (2006)
	Fipronil	Tang et al. (2004)
	Malathion	Buratti et al. (2006)
	Parathion	Mutch et al. (2003); Mutch and Williams, (2006); Buratti et al. (2006); Foxenberg et al. (2007)
	Phorate	Usmani et al. (2004b)
	Sulprofos	Usmani et al. (2004b)
	Deltamethrin	Godin et al. (2007)
	Esfenvalerate	Godin et al. (2007)
	Methoxychlor	Hu and Kupfer (2002)
	Insect repellent: DEET	Usmani et al. (2002)
CYP3A6	*Insecticide:* fipronil	Tang et al. (2004)

(Continued)

Table 5.3 (Continued)

Phase I isoform	Substrate	Reference
CYP3A7	*Insecticides*	
	Carbofuran	Usmani et al. (2004a)
	Chlorpyrifos	Buratti et al. (2006); Foxenberg et al. (2007)
	Endosulfan-α	Casabar et al. (2006)
	Fenthion	Buratti et al. (2006)
	Fipronil	Tang et al. (2004)
	Malathion	Buratti et al. (2006)
	Parathion	Foxenberg et al. (2007)
	Herbicide: atrazine	Joo et al. (2010)
FMO1	*Insecticides*	
	Aldicarb	Schlenk et al. (2002)
	Demeton-O	Furnes and Schlenk (2005)
	Disulfoton	Usmani et al. (2004b)
	Ethiofencarb	Furnes and Schlenk (2005)
	Fonofos	Furnes and Schlenk (2005)
	Methiocarb	Furnes and Schlenk (2005); Usmani et al. (2004b)
	Sulprofos	Usmani et al. (2004b)
	Phorate	Hodgson et al. (1998); Usmani et al. (2004b)
FMO2.1	*Insecticides*	
	Disulfoton	Henderson et al. (2004)
	Phorate	Henderson et al. (2004)
FMO3	*Insecticides*	
	Aldicarb	Schlenk et al. (2002)
	Demeton-O	Furnes and Schlenk (2005)
	Ethiofencarb	Furnes and Schlenk (2005)
	Fenthion	Furnes and Schlenk (2004); Leoni et al. (2008)
	Fonofos	Furnes and Schlenk (2005)
FMO5	*Insecticide:* fenthion	Furnes and Schlenk (2004); Leoni et al. (2008)
Carboxylesterase	*Insecticides*	
	Bioresmethrin	Ross et al. (2006)
	Permethrin	Crow et al. (2007); Ross et al. (2006)
	Pyrethroid-like model substrates	Huang et al. (2005)
	Malathion	Buratti and Testai (2005)
Carboxylesterase CE1	*Insecticides*	
	Cypermethrin analogs	Huang et al. (2005)
	Deltamethrin	Godin et al. (2006)
	Esfenvalerate	Godin et al. (2006)
	Fenvalerate analogs	Huang et al. (2005)
Carboxylesterase CE2	*Insecticides*	
	Deltamethrin	Godin et al. (2006)
	Esfenvalerate	Godin et al. (2006)
PON1-192	*Insecticides*	
	Chlorpyrifos oxon	Mutch et al. (2007); Richter et al. (2009); Richter and Furlong (1999)

(Continued)

Table 5.3 (Continued)

Phase I isoform	Substrate	Reference
	Diazoxon	Mutch et al. (2007); Richter et al. (2009); Richter and Furlong (1999)
	Paraoxon	Mutch et al. (2007); Richter et al. (2009); Richter and Furlong (1999)

(Anand et al., 2006a,b; Crow et al., 2007; Godin et al., 2006, 2007; Huang et al., 2005; Price et al., 2008; Ross and Crow, 2007; Ross et al., 2006; Wheelock et al., 2005) and humans (Choi et al., 2002; Crow et al., 2007; Godin et al., 2006, 2007; Huang et al., 2005; Price et al., 2008; Ross and Crow, 2007; Ross et al., 2006), as well as the role of carboxylases in the human hepatic metabolism of malathion (Buratti and Testai, 2005) (see Tables 5.1 and 5.3). The phase I metabolism of methoxychlor, because of its importance as an environmental endocrine disruptor, has received considerable attention, particularly in the laboratory of the late David Kupfer (e.g., Bikadi and Hazai, 2008; Hazai et al., 2004; Hu and Kupfer, 2002a,b).

Other recent studies of hepatic pesticide metabolism include those on azole fungicides (Barton et al., 2006; Mazur et al., 2007), the carbamate insecticide terbutol (Suzuki et al., 2001), and the herbicide diuron (Abass et al., 2007a); see Tables 5.1 and 5.3.

Phase II Enzymes

Phase II conjugation reactions of pesticides are less well known than phase I, although several types of conjugation are known to involve pesticides as substrates (Dorough, 1984; Matsumura, 1985; Mehendale and Dorough, 1972; Motoyama and Dauterman, 1980). Glucuronides are important metabolites of carbamates, including banol, carbaryl, and carbofuran (Mehendale and Dorough, 1972), as well as the endocrine disruptor methoxychlor (Hazai et al., 2004), and some organophosphorus and other pesticides (Hutson, 1981). Ethereal sulfates, although not important in pesticide metabolism, may be formed from the oxidative metabolites of carbaryl and carbofuran (Dorough, 1968, 1970). GST is important in the metabolism of organophosphorus pesticides (Abel et al., 2004a; Choi et al., 2006; Motoyama and Dauterman, 1980) and halogenated herbicides such as the chloroacetanilides and chloro-S-triazines (Abel et al., 2004; Cho and Kong, 2007) as well as thiocarbamate herbicides such as molinate (Campbell et al. 2008). The conjugation products are typically further metabolized and, in humans, excreted as urinary mercapturic acids. Interestingly, the addition of GSH, with a molecular weight of 307, to these 200- to 300-molecular-weight xenobiotics creates a product that exhibits a species-dependent disposition due to species differences in size thresholds for biliary excretion compared to renal excretion.

Although not strictly speaking a detoxication reaction, hepatic aliesterase, by forming a stable phosphorylated enzyme with organophosphates (oxons), may serve as an inert storage protein (Chambers et al., 1990).

Comparative Aspects

Limitations of time and space militate against a comparative approach to pesticide metabolism and this chapter is devoted, with only a few exceptions, to this subject as it applies to humans and to surrogate mammals used in research and risk analysis.

However, pesticide metabolism in fish liver is of increasing importance to environmental toxicologists and has been the subject of many recent reports. The diversity of these reports is illustrated by the following examples: *Carassius auratus* and alachlor (Yi et al., 2007), *Gasterosteus aculeatus* and prochloraz (Sanchez et al., 2008), *Ictalurus punctatus* and methoxychlor (James et al., 2008), *Micropterus salmoides* and p,p-DDE (Barberm et al., 2007), *Oncorhynchus mykiss* and dieldrin (Barnhill et al., 2003), *Oreochromis mossambicus* and monocrotophos (Rao, 2006), *Oreochromis niloticus* and paraquat (Figueiredo-Fernandes et al., 2006), *Rhamdia quelen* and glyphosate (Glusczak et al., 2007), and *Salmo salar* and p,p-DDE (Mortensen and Arukwe, 2006). There is a much smaller body of literature on birds (e.g., Cortright and Craigmill, 2006), reptiles (e.g., Gunderson et al., 2006), and various food and feral mammals (e.g., Dupuy et al., 2001).

Biotransformation in Extrahepatic Tissues

The liver is generally more important than other organs in the biotransformation of xenobiotics, including pesticides. However, other organs and tissues may be active to some degree. For example, it was shown early that DDT is degraded by rat diaphragm, kidney, and brain in vitro (Judah, 1949). Later study showed that these changes proceeded at a very slow rate in vivo. However, as shown in the following sections, not all extrahepatic metabolism is inefficient.

Some enzymes outside the liver may be induced, but the matter has received little attention. Wattenberg (1971) demonstrated that the small intestines of rats fed a balanced purified diet or starved for 1 day possess virtually no benzo[*a*]pyrene hydroxylase activity, whereas the intestines of rats fed the same diet plus turnip greens, broccoli, cabbage, or brussels sprouts have marked activity of this enzyme. The same activity in human skin is induced by polycyclic hydrocarbons (Alvares et al., 1973). Neal (1972) showed that monooxygenases of the lung active in the metabolism of parathion can be induced by phenobarbital. Induction of enzymes that metabolize pesticides is considered in detail in Chapter 7.

Lung

The lung is a primary site of exposure to airborne as well as blood-borne environmental pollutants, such as pesticides, and for this reason it is a target organ for many chemically induced toxicities (Bond, 1983, 1993; Dahl and Lewis, 1993; Ding and Kamienski, 2003). Because the lung has a full complement of metabolic enzymes, it has the capacity to activate and deactivate pesticides and other xenobiotics. Early studies showed that parathion is metabolized to paraoxon and diethylphosphorothioic acid by rabbit lung at about 20% of the rate in liver (Neal, 1972).

Several studies have demonstrated the importance of pulmonary CYP and FMO enzymes in pesticide oxidation (Feng et al., 1990; Li et al., 1992). In the lung, FMO appears to play a more important role than CYP in the oxidation of certain pesticides and xenobiotics (Kinsler et al., 1988; Tynes and Hodgson, 1983, 1985a,b). Other studies have shown the existence of an FMO form now known as FMO2 in the lung that is not present in the liver (Lawton et al., 1990; Tynes and Hodgson, 1983; Tynes et al., 1985; Venkatesh et al., 1992b; Williams et al., 1984, 1985). Boland et al. (2004) demonstrated the metabolism of naphthalene, primarily to the dihydrodiol, in respiratory tissues of rhesus monkeys, and the metabolism and toxic effects of naphthalene in respiratory tissues continues to be of interest (Bogen et al., 2008).

Nasal Tissues

The nasal mucosa is the first tissue of contact for inhaled xenobiotics and compounds have been identified that cause nasal lesions or tumors in experimental animals.

The drug-metabolizing activity of nasal tissues has been reviewed by Reed (1993) and, more recently, by Ding and Kamienski (2003). Enzymes known to be present include a variety of CYPs (CYP1A1, 2B1, 2E1, 3A1, 4A1, 2G1), FMOs, carboxylesterases, epoxide hydrolases, glutathione S-transferases, and UDP-glucuronyl transferases. It is of some interest that, despite the low concentrations of nasal CYP enzymes, these have been demonstrated to have greater specific activity toward several substrates than liver CYPs, perhaps as a result of higher ratios of NADPH cytochrome P450 reductase to CYP in the nasal tissues. Nasal CYPs appear to be less inducible than liver isoforms, although they appear to be sensitive to a number of CYP inhibitors.

Few pesticides are known to give rise to toxic endpoints in the nasal tissues. However, alachlor, a restricted-use chloroacetamide herbicide that at one time was widely used in agriculture, was demonstrated to cause rare nasal carcinomas in rats. The putative metabolic product thought to be responsible for its carcinogenicity was identified as diethylbenzoquinoneimine (DEBQI), which is produced only after extensive metabolism of alachlor, involving CYPs as well as an aryl amidase. Human CYP isoforms 2B6 and 3A4 are among those that have been identified as being important in the production of metabolite precursors to DEBQI (Coleman et al., 1999, 2000).

Genter and co-workers (Deamer et al., 1994; Genter et al., 1995, 1998) have demonstrated the role of microsomal epoxide hydrolase and CYP2E1 in the nasal toxicity of dichlobenil in the mouse. It was subsequently shown that CYP2A10 and 2A11, isoforms that comprise some 25% of the total olfactory CYP content, also play an important role in the nasal toxicity of dichlobenil (Ding et al., 1994, 1996).

Skin

Because the skin is the largest organ in the human body, is continuous over the surface area of the body, and is in direct contact with the environment, it is often the portal of entry for pesticides and other xenobiotics. The skin is known to contain many of the XMEs found in the liver, and some of these have been shown to be inducible, primarily by polycyclic hydrocarbons (Goerz et al., 1994; Jugert et al., 1994; Baron et al., 2008). The metabolic capacity of skin for pesticides was shown early when slices of rabbit skin were shown to hydrolyze paraoxon at a concentration of $7.7 \times 10^{-3}\,M$ to the extent of 20% in 1 h/g of tissue. Because the absorption of paraoxon and related compounds is slow, this metabolism may be an important defense mechanism (Fredriksson et al., 1961). By use of in vitro methods, such as the isolated perfused porcine skin flap (Carver et al., 1990) and mouse skin microsomes (Venkatesh et al., 1992a), the skin has been shown to have the capacity to metabolize a variety of pesticides. For example, Chang et al. (1994), using the isolated perfused porcine skin flap, showed that both carbaryl and parathion were metabolized during uptake by the skin.

Kidney

Because of the kidney's high blood flow, its ability to concentrate chemicals, and the presence of renal XMEs, the kidney may also be a site of toxicity from xenobiotics. Many of these toxic effects can be directly attributable to the presence and localization of specific forms of enzymes responsible for activation (Hu et al., 1993; Speerschneider and Dekant, 1995). Several studies have highlighted the importance of renal oxidative enzymes, particularly FMO, in the metabolism of pesticides and other xenobiotics (Kinsler et al., 1988; Tynes and Hodgson, 1983). As was the case with the lung, the renal FMO enzymes played a greater role in microsomal systems in the oxidation of several pesticides than renal CYP, suggesting an important role for FMO in the extrahepatic metabolism of toxicants. Studies of kidney FMO have provided evidence for several isoforms in the kidney, including the forms found in liver and lung (Atta-Asafo-Adjei et al., 1993; Burnett et al., 1994; Ripp et al., 1999; Venkatesh et al., 1991), and Furnes and Schlenk (2005) have demonstrated the sulfoxidation of fenthion and methiocarb by kidney FMO.

Central Nervous System

Very little is known about XMEs in the central nervous system (CNS). Several studies have demonstrated CYP activity and constitutive expression of various CYP isozymes (Britto and Wedlund, 1992; Ghersi-Egea et al., 1993; Hansson et al., 1992; Hodgson et al., 1993; Miksys and Tyndale, 2009). The activation or detoxication of pesticides by the CNS is of particular interest in the case of pesticides that both exhibit their action and are metabolized in the brain. Studies by Chambers and Chambers (1989) demonstrated that the neurotoxicity of a series of organophosphorus compounds correlated better with their activation in the brain than with activation in the liver.

Several studies have reported activity of known FMO substrates by brain microsomes (Bhamre et al., 1993; Duffel and Gillespie, 1984; Kawaji et al., 1994), and one form of FMO has been demonstrated using polymerase chain reaction amplification (Blake et al., 1996).

Gastrointestinal Tract

Xenobiotic metabolism in the gastrointestinal tract has been reviewed by Ding and Kamienski (2003). Some carbaryl is hydrolyzed and the resulting naphthol is conjugated with glucuronic acid by the intestine (Pekas and Paulson, 1970), and Furnes and Schlenk (2005) have demonstrated sulfoxidation of fenthion by FMO in the intestine.

METABOLISM IN HUMANS

For several reasons, pesticide metabolism in humans is a rapidly increasing area of research. Since approximately 2000, human hepatocytes, human cell lines, human cell fractions, and recombinant human xenobiotic-metabolizing enzymes have become increasingly available and, as a result, ethical experiments can be carried out without risk. Prior to that time pesticide metabolism was studied in surrogate animals, primarily rodents, and the results, for regulatory purposes, were extrapolated to humans. Utilizing human materials directly avoids this extrapolation and, at the least, helps in selecting the most appropriate surrogate animal. Of equal importance is the investigation of variation in the human population, the identification of subgroups or individuals in the population potentially at greater risk, and the identification of human-specific interactions; things that cannot be done in surrogate animals.

Much of the variation in xenobiotic metabolism in humans is due to the occurrence of most, if not all, xenobiotic-metabolizing enzymes in a number of polymorphic forms. A polymorphism is defined as an inherited monogenetic trait that exists in the population in at least two genotypes (two or more variant alleles) and is stably inherited. Several modified alleles may occur at the same gene locus, and population differences in the incidence of polymorphisms are known to occur in pesticide-metabolizing CYPs (Hodgson and Croom, 2008). Single-nucleotide polymorphisms and their effects have been intensively studied in the metabolism of clinical drugs but there are, as yet, few such studies on the metabolism of pesticides. To date these studies have shown the importance of polymorphic forms in the metabolism of carbaryl, carbofuran and sulprofos by CYP2C9 and chlorpyrifos by CYP2B6, CYP2C19, and CYP3A4 (Dai et al., 2001; Tang et al., 2001, 2002; Usmani et al., 2004a,b; Rose et al., 2005; Croom et al., 2010).

The human metabolism of pesticides is included in the preceding sections and some examples are given in Table 5.1. However, because of the important and topical nature of this subject, the enzymes involved in human phase I metabolism of pesticides are summarized in Table 5.3. A database—In Vitro Human Metabolism of Agrochemicals and Related Chemicals (Hodgson, 2011)—is available through the Foundation for Toxicology and Agromedicine (www.toxicologyagromed.org).

TOXICITY OF METABOLITES

In general, metabolites are less toxic than their parent compounds, if for no other reason than that they are usually more water soluble and, therefore, more rapidly excreted. There are notable exceptions in which biotransformation results in an inherently more toxic product. Such reactions are generally referred to as activation reactions. These reactive metabolites may combine covalently with cellular constituents such as DNA, RNA, or proteins, and carcinogenesis, mutagenesis, and cellular necrosis are often attributable to such reactive metabolites (Anders et al., 1992; Guengerich, 1992, 1993; Levi and Hodgson, 2001; Parke, 1987). Hollingworth et al. (1995) reviewed reactive metabolites with particular reference to agrochemicals,

The metabolic production of a more toxic compound is sometimes called lethal synthesis to emphasize that biotransformation in this instance is the source of danger. The term lethal synthesis was introduced in a lecture given on June 7, 1951, by Peters (1952) in connection with fluoroacetic acid. This compound is not itself an enzyme inhibitor, but is converted by enzymes into a highly toxic material. Peters (1963) later reviewed and extended the concept of lethal synthesis, although the term itself is now seldom used.

The effects of metabolism may be complex, as illustrated by studies of bromobenzene. It has been known for some time that the liver necrosis associated with this compound is caused by one or more toxic metabolites. Stimulation of its biotransformation by phenobarbital potentiates the injury of toxic doses to the liver, and inhibition of its metabolism by SKF 525-A prevents this injury. However, although 3-methylcholanthrene causes a slight in vitro stimulation of the metabolism of bromobenzene and does not alter the overall rate in vivo, it does protect against the hepatotoxicity. Rats dosed with bromobenzene after induction with 3-methylcholanthrene excrete more bromophenyldihydrodiol, bromocatechol, and 2-bromophenol than do uninduced rats. The increase in the first two compounds suggests an increased capacity to detoxify the highly reactive epoxide. The increase in 2-bromophenol suggests that induction by 3-methylcholanthrene diverts the metabolism of bromobenzene to a comparatively nontoxic pathway (Zampaglione et al., 1973).

The role of metabolic activation in the carcinogenic process, particularly in the formation of DNA adducts by reactive metabolites, has been of particular concern. For example, monooxygenase enzymes have been postulated to play a role in the metabolic activation of alachlor and metolachlor (Brown et al., 1988; Feng and Wratten, 1989; Feng et al., 1990; Jacobsen et al., 1991; Li et al., 1992). Although studies suggest that alachlor has a greater carcinogenic potential than metolachlor, the carcinogenic responses to these compounds are species- and tissue-specific, alachlor being a nasal-specific carcinogen in rats but not in mice. Metolachlor, on the other hand, is carcinogenic to the liver but not to nasal tissue (U.S. Environmental Protection Agency, 1986, 1987). Available evidence suggests that the species- and tissue-specific responses observed, particularly for alachlor, result from specific metabolic enzymes, including monooxygenases and arylamidases, and the generation of the putative carcinogenic metabolite, diethylbenzoquinone imine (see Coleman et al., 1999, 2000, for appropriate references).

The most studied generation of reactive metabolites from pesticides is the generation of oxons from organophosphorus compounds containing the $P=S$ moiety by oxidative desulfuration. Not only does this reaction produce the oxons, cholinesterase inhibitors responsible for the neurotoxicity of these compounds, but it also releases reactive sulfur, a potent CYP inhibitor (Neal, 1980; Neal and Halpert, 1982; Neal et al., 1983); see Chapter 7 for further information on inhibition. The hepatic metabolism of organophosphorus insecticides continues to be investigated in both humans and rodents. In general, through a common intermediate containing a phosphiithrane ring, they are either activated to their oxons, potent inhibitors of acetylcholinesterase and other esterases, or detoxified. This intermediate is generated by CYP isoforms, particularly CYP1A2; CYP2B6, and CYP 3A4 (see Tables 5.1 and 5.3).

The mode of action of the insecticide synergist piperonyl butoxide and other methylenedioxyphenyl compounds is also due to a reactive metabolite, believed to be a carbene derivative, that combines with the heme iron of CYP (Dahl and Hodgson, 1979; see Chapter 7).

PHYSIOLOGICAL FACTORS AFFECTING BIOTRANSFORMATION

Species, strains, and individuals may all vary in their susceptibility to toxicants, including pesticides. In some cases it has been possible to explain these differences by one of several causes, including differences in metabolism. In this section, some examples are presented in which it is known that the activity of microsomal enzymes is influenced by age, gender, and species.

Developmental Effects

Microsomal XME activity is low in the fetus and the newborn, but increases rapidly during the early days or weeks of life (Croom et al., 2009; Fouts and Adamson, 1959; Ronis and Cunny, 1994, 2000). For this reason fetuses and newborns are often more susceptible to certain drugs and xenobiotics than adults. With aging, there is generally a decrease in enzymatic activity, although increases in some activities have been observed (Kitahara et al., 1982; Van Bezooijen, 1984; Van Bezooijen et al., 1986).

In the mouse, FMO1 and FMO5 are expressed as early as gestation days 15 and 17, and equally between genders until puberty. FMO3 is not expressed until 2 weeks postpartum and is found equally in males and females until 6 weeks postpartum, when it becomes undetectable in the male. This developmental pattern, as seen in the female mouse, is similar to that seen in humans of either gender (Cherrington et al., 1998a).

Species Differences

Many species differences in the metabolism of xenobiotics can be explained in terms of differences in the activity of liver microsomal enzymes (Brodie and Maickel, 1962; Quinn et al., 1958; Walker, 1994), although broad similarities often exist across large systematic groups. For example, Barron et al. (1993) observed that the detoxication of chlorpyrifos in channel catfish was similar to that in other vertebrates in both phase I and phase II metabolism. An example of a comparative study of a specific pesticide is that of Chin et al. (1979b) on carbaryl. The gender-dependent expression of FMO iso-forms also varies between species, as outlined below.

Individual and Strain Differences

Strain and individual differences are often discussed conveniently in terms of tolerance and resistance, both implying reduced susceptibility to a toxicant. The word *tolerance* is used when the observed decrease in susceptibility occurs in an individual organism as a result of its own previous or continuing exposure to the particular toxicant or to some other conditioning stimulus, whereas *resistance* refers to a change in a population brought about by genetic selection.

Differences in basic levels of enzyme activity have been detected not only among different species, but also among different strains of mice (Jay, 1955), rats (Quinn et al., 1958), rabbits (Cram et al., 1965), and birds (Ronis and Walker, 1989; Walker, 1983). Comparative aspects of xenobiotic metabolism, particularly as they relate to CYPs, were reviewed by Hodgson (1979) and Kulkarni et al. (1975).

Gender Differences

Metabolism of xenobiotics may vary with the gender of the organism, and in some cases differences in overall toxicity between males and females of various species are known (Bonate, 1991). In the absence of induction, microsomal enzyme activity is often higher in the adult male rat than in females or immature males. However, the stimulatory effect of xenobiotics on microsomal enzymes is usually greater in females and immature males than in the adult male rat (Conney and Burns, 1962). Gender differences become apparent at puberty and are usually maintained throughout adult life. The differences in microsomal monooxygenase activity between males and females have been shown to be under the control of sex hormones, at least in some species. Sexually dimorphic CYPs appear to arise, in the rat, by programming, or imprinting, that occurs in neonatal development. This imprinting is brought about by a surge of testosterone that occurs in the male neonate and appears to imprint the developing hypothalamus, so that in later development growth hormone is secreted in a gender-specific manner. This pattern of growth hormone production is pulsatile and the higher level of circulating testosterone in the male maintains the expression of male-specific CYP isoforms such as 2C11. On the other hand, a more continuous pattern of growth hormone secretion and the lack of circulating testosterone appear to be responsible for female-specific CYPs such as 2C12 (Gonzalez, 1989; Hosteter et al., 1987; Kobliakov et al., 1991; Schenkman et al., 1989).

Gender-specific expression is also seen with the FMO enzymes. It has been known for some time that hepatic FMO activity is higher in female mice than in males, and that the lower levels in males result from testosterone repression (Dannan et al., 1986; Duffel et al., 1981; Falls et al., 1997; Lemoine et al., 1991; Wirth and Thorgeirsson, 1978). In addition, hormonal changes during pregnancy have been reported to increase FMO levels (Williams et al., 1985). With regard to pesticide oxidation, gender differences have also been observed, with higher activity in female mouse liver than in male (Kinsler et al., 1988).

Recent studies have identified FMO isozymes involved in these gender differences and some of the hormonal factors involved in regulation. In several strains of mouse liver, FMO1 expression was found to be two to three times higher in female mice compared to males, and FMO3, expressed in females at levels comparable to FMO1, was not detectable in male liver (Falls et al., 1995, 1997; Cherrington et al., 1998a). In rat liver, however, FMO1 is higher in the male, whereas FMO3 is gender-independent in both rats and humans. FMO5 is gender-independent for the mouse, rat, and human (Cherrington et al., 1998a).

Genetic Factors

The existence of discontinuous or biphasic variation is a strong indication of the possibility of genetic involvement. Some examples are given here in which such variation in enzyme activity within populations has been proven to be genetic in origin.

The fungicide ziram caused hemolytic anemia with Heinz body formation in a man later shown to be deficient in erythrocyte glucose-6-phosphate dehydrogenase. Ziram also caused one of the typical in vitro reactions, formation of Heinz bodies, in the blood of another person known to be deficient in this enzyme (Pinkhas et al., 1963).

Many of the phenotypic variations in drug responses observed in human populations have been shown to result from polymorphisms in the expression of the xenobiotic-metabolizing enzymes (for reviews see Smith et al., 1994a,b; Kalow, 1991; Coutts and Urichuk, 1999; and Wormhoudt et al., 1999). Many human CYP isoforms have been shown to be polymorphic (Daly et al., 1998; Goldstein and De Morais, 1994; Smith et al., 1994a,b). Pesticides metabolized by these polymorphic CYP isoforms, such as chlorpyrifos by CYP2B6 and CYP3A4, could have higher or lower risk factors in some proportion of the exposed population dependent upon the distribution of the relevant polymorphisms (Dai et al., 2001; Hodgson and Rose, 2007; Tang et al., 2001).

TOLERANCE AND RESISTANCE

The terms resistance and tolerance refer to a relative lack of susceptibility of a population of organisms to the effects of a toxicant. If the genetic trait pre-exists in a population so that it is obvious when the population is first exposed to the toxicant, this should be regarded as tolerance rather than resistance, the latter term being reserved for those cases in which the trait is brought to an observable level only through selection brought about by exposure to the toxicant. It is clear that many instances of resistance are based on differences in toxicant metabolism that distinguish the resistant from the nonresistant population.

Tolerance

Tolerance to a compound is often the result of an organism's increased ability to metabolize the chemical subsequent to an initial exposure. This is true, for example, in connection with the pesticides nicotine (Werle and Uschold, 1948) and dieldrin (Wright et al., 1972).

In a few instances, it has been shown clearly that the increased metabolism responsible for tolerance was mediated by higher activity of the microsomal enzymes of the liver. It seems likely that the same explanation will hold in connection with some other instances of tolerance. As recorded in Chapter 7, pesticides frequently act as inducers of microsomal enzymes. Because activity of these enzymes usually leads to detoxication, it seems likely that many of the compounds listed as inducers are capable of producing tolerance under suitable conditions.

Tolerance also may exist in situations in which it has been impossible to demonstrate any increased ability to metabolize the toxicant; finally, there are instances of tolerance for which the mechanism is not only unknown but unexplored. For example, rodents may develop true tolerance, as distinguished from bait shyness, to a number of rodenticides including arsenic oxide, zinc phosphide, strychnine, sodium fluoroacetate, ANTU, and norbormide (Lund, 1964, 1967). Certain populations of pine mice subjected to control with endrin lost susceptibility to the compound, but sublethal exposure conferred a degree of tolerance regardless of the past history of the population (Webb and Horsfall, 1967).

Resistance

Resistance in the toxicological sense is better known in insects and a variety of other pest species than in vertebrates. Many insect species with public health importance have long been known to be resistant to one or more pesticides (Brown and Pal, 1971; Georghiou and Saito, 1983) and a much larger number of agricultural pests are also resistant. The numbers identified in both groups continue to grow. Resistance to a particular compound does not involve an entire species but only the toxicant-stressed population; nevertheless, resistance constitutes a serious public health and economic problem.

Resistant strains are first recognized only after the parent population has been selected by killing off many of its susceptible members. The rate at which selection progresses depends not only on the intensity of the selection pressure, but also on the duration of each generation. Therefore, it is not surprising that resistance has most often been observed among organisms such as bacteria or houseflies, characterized by large numbers of individuals and a rapid rate of reproduction. However, resistance has also been observed in species with relatively small populations and relatively slow multiplication.

Resistance of a vertebrate species to a pesticide apparently was first recognized in the 1960s and 1970s in connection with Norway rats exposed to warfarin. This phenomenon was first reported from Scotland (Boyle, 1960) and subsequently from Denmark (Lund, 1964, 1967), England and Wales (Bentley, 1969; Drummond, 1966), The Netherlands (Ophof and Langeveld, 1969), Germany (Telle, 1971), and the United States, specifically North Carolina and Idaho (Brothers, 1972; Jackson and Kaukeinen, 1972). Resistance in rats continues to be reported and investigated on a regular basis, for example, in France (Lasseur et al., 2005), in Germany (Pelz, 2007), and in China (Wang et al., 2008).

The early literature on the resistance of mammals to warfarin was reviewed by Lund (1967). Only a few points need to be recorded here. So far, resistance is known to occur in four species; the Norway and roof rats, the house mouse, and humans. In addition, in their original studies of coumarin compounds, Link and his students reported marked variation in susceptibility in rabbits as a Mendelian characteristic (Campbell et al., 1941). The exact mechanism of inheritance of resistance to warfarin is not clear. In humans, the facts are consistent with transmission by a single autosomal dominant gene (O'Reilly et al., 1963), but this was based on only one kindred. In rats and mice, it seems that more than one gene is involved. The physiological basis of the resistance in vertebrates also is not clear and may be different in different instances. It is generally held that it involves the vitamin K cycle (Cain et al., 1998) and polymorphisms in the VKOR vitamin K epoxide reductase gene (Lasseur et al., 2005; Wang et al., 2008). The role of CYP isoforms in anticoagulant resistance is not clear, although it has been noted in bromadiolone resistance in the rat that several CYP isoforms are overexpressed (Markussen et al., 2008a,b). In every instance studied, including humans, resistance extended to other coumarin anticoagulants and those based on indanedione, while susceptibility to heparin is normal.

Another early report of resistance among vertebrates involved mosquito fish collected from insecticide-contaminated waters near cotton fields (Vinson et al., 1963).

Further study revealed 2- to 500-fold levels of resistance to a variety of pesticides in mosquito fish and five other species of fish (Boyd and Ferguson, 1964a,b; Ferguson and Bingham, 1966a,b; Ferguson and Boyd, 1964; Ferguson et al., 1964, 1965). Resistance to chlorinated hydrocarbon pesticides was found in three species of frog (Boyd et al., 1963). The degree of resistance may be so great in some instances that resistant species can withstand enough poison to kill their predators.

Ozburn and Morrison (1962) were the first to produce resistance to a pesticide in a mammal by selection under laboratory conditions. In mice selected by a single intraperitoneal dose of DDT administered at 4 weeks of age, resistance in the ninth generation had increased by a factor of 1.7 as measured by the LD_{50}. Although the factor of 1.7 is small, about half of the susceptible mice withstood a dose that was uniformly fatal to control mice. Further study (Ozburn and Morrison, 1965) revealed that the selected and control colonies differed in their rates of oxygen consumption. The resistant mice were fatter than the susceptible ones, and considerable evidence indicated that resistance depended on preferential deposit of DDT in the fat and consequently the avoidance of peak levels in sensitive tissues. The resistance was not specific to DDT but extended to lindane and dieldrin (Barker and Morrison, 1966). Success in the development of resistance in mammals in the laboratory has not been uniform; apparently some strains are not sufficiently heterozygous to respond to selection (Guthrie et al., 1971).

Thus many instances are known in which species or strains differ in their susceptibility to pesticides, the resistance arising through selection in the field. In other instances, it has been possible to produce resistance in the laboratory through selection. In many instances it has been possible to define the genetic mechanisms responsible for observed differences in the metabolism of the pesticides in question. Except in the case of insects, a genetic mechanism has been defined only rarely in connection with metabolism of pesticides.

The following references are suggested on this subject: Dauterman (1994), Evered and Collins (1984), Georghiou and Saito (1983), and Hayes et al. (1990).

CONCLUSIONS

Knowledge of the metabolism of pesticides is essential for several reasons, including the development of more selective insecticides, and provides, in part, the fundamental basis for science-based risk assessments for human and environmental health (Buratti et al., 2007; Hodgson and Rose, 2005, 2007). Until recently, and as a matter of necessity, this research was carried out almost exclusively on experimental animals and the results, particularly in the case of human health risk assessments, were extrapolated to humans. Although much essential background will continue to be obtained from experimental animals, because of the ready availability of human hepatocytes, human cell lines, human cell fractions, and recombinant human enzymes, essential information will, in the future, be derived to a greater extent directly from human studies. At the same time, studies utilizing surrogate animals will also be revolutionized by such new techniques of molecular biology as the use of knockout and transgenic, including "humanized", mice (e.g., Gonzalez, 2003), and the knowledge of the genomes of many species. These same techniques through the study of genetic polymorphisms will enable us to identify human populations at increased risk and enable comparative studies to be carried out at the level of specific isoforms of the XMEs involved. Thus the study of pesticide metabolism continues to evolve into a new, more molecular, era that will be fascinating as well as useful.

REFERENCES

Abass, K., Reponen, P., Turpeinen, M., Jalonen, J., & Pelkonen, O. (2007a). Characterization of diuron N-demethylation by mammalian hepatic microsomes and cDNA-expressed human cytochrome P450 enzymes. *Drug Metab. Dispos.*, *35*, 1634–1641.

Abass, K., Reponen, P., Jalonen, J., & Pelkonen, O. (2007b). In vitro metabolism and interaction of profenofos by human, mouse and rat liver preparations. *Pestic. Biochem. Physiol.*, *87*, 238–247.

Abass, K., Reponen, P., Jalonen, J., & Pelkonen, O. (2007c). In vitro metabolism and interactions of the fungicide metalaxyl in human liver preparations. *Environ. Toxicol. Pharmacol.*, *23*, 39–47.

Abass, K., Reponen, P., Mattila, S., & Pelkonen, O. (2009). Metabolism of carbosulfan. I. Species differences in the in vitro biotransformation by mammalian hepatic microsomes, including human. *Chem. Biol. Interact.*, *181*, 210–219.

Abass, K., Reponen, P., Mattila, S., & Pelkonen, O. (2010). Metabolism of carbosulfan. II. Human interindividual variability in its in vitro hepatic biotransformation and the identification of the cytochrome P450 isoforms involved. *Chem. Biol. Interact.*, *185*, 163–173.

Abel, E. L., Opp, S. M., Verlinde, C. L., Bammler, T. K., & Eaton, DL. (2004a). Characterization of atrazine biotransformation by human and murine glutathione S-transferases. *Toxicol. Sci.*, *80*, 230–238.

Abel, E. L., Bammler, T. K., & Eaton, D. L. (2004b). Biotransformation of methyl parathion by glutathione S-transferases. *Toxicol. Sci.*, *79*, 224–232.

Adams, N. H., Levi, P. E., & Hodgson, E. (1990). In vitro studies of the metabolism of atrazine, simazine and terbutryn in several mammalian species. *J. Agric. Food Chem.*, *38*, 1411–1417.

Ahmed, M. K., Casida, J. E., & Nichols, R. E. (1958). Bovine metabolism of organophosphorus insecticides: Significance of rumen fluid with particular reference to parathion. *J. Agric. Food Chem.*, *6*, 740–746.

Aitio, A. (1978). A simple and sensitive assay of 7-ethoxycoumarin deethylation. *Anal. Biochem.*, *85*, 488–491.

Akhtar, M. H. (1984). Metabolism of deltamethrin by cow and chicken liver enzyme preparation. *J. Agric. Food Chem.*, *32*, 258–262.

Alvares, A. P., Kappas, A., Levin, W., & Conney, A. H. (1973). Inducibility of benzo[*a*]pyrene hydroxylase in human skin by polycyclic hydrocarbons. *Clin. Pharmacol. Ther.*, *14*, 30–40.

Anand, S. S., Bruckner, J. V., Haines, W. T., Muralidhara, S., Fisher, J. W., & Padilla, S. (2006). Characterization of deltamethrin metabolism by rat plasma and liver microsomes. *Toxicol. Appl. Pharmacol.*, *212*, 156–166.

Anand, S. S., Kim, K. B., Padilla, S., Muralidhara, S., Kim, H. J., Fisher, J. W., et al. (2006). Ontogeny of hepatic and plasma metabolism of deltamethrin in age-dependent neurotoxicity. *Drug Metab. Dispos.*, *34*, 389–397.

Anders, M. W., Dekant, W., & Vamvakas, S. (1992). Glutathione-dependent toxicity. *Xenobiotica*, *22*, 1135–1145.

Atta-Asafo-Adjei, E., Lawton, M. P., & Philpot, R. M. (1993). Cloning, sequencing, distribution, and expression of a mammalian flavin-containing monooxygenase from a third gene subfamily. *J. Biol. Chem.*, *268*, 9681–9689.

Baldwin, M. K., Robinson, J., & Parke, D.V. (1972). A comparison of the metabolism of HEOD dieldrin in the CFI mouse with that in the CFE rat. *Food Cosmet. Toxicol.*, *10*, 333–351.

Barberm, D. S., McNally, A. J., Garcia-Reyero, N., & Denslow, N. D. (2007). Exposure to p,p′-DDE or dieldrin during the reproductive season alters hepatic CYP expression in largemouth bass Micropterus salmoides. *Aquat. Toxicol.*, *81*, 27–35.

Barker, P. S., & Morrison, F. O. (1966). The basis of DDT tolerance in the laboratory mouse. *Can. J. Zool.*, *44*, 879–887.

Barnhill, M. L., Rosemond, M.V. M., & Curtis, L. R. (2003). Dieldrin stimulates biliary excretion of C-14-benzoapyrene polar metabolites but does not change the biliary metabolite profile in rainbow trout Oncorhynchus mykiss. *Toxicol. Sci.*, (75), 249–258.

Baron, J. M., Wiederholt, T., Heise, R., Merk, H. F., & Bickers, D. R. (2008). Expression and function of cytochrome P450-dependent enzymes in human skin cells. *Curr. Med. Chem.*, *15*, 2258–2264.

Barron, M. G., Plakas, S. M., Wilga, P. C., & Ball, T. (1993). Absorption, tissue distribution and metabolism of chlorpyrifos in channel catfish following waterborne exposure. *Environ. Toxicol. Chem.*, *12*, 1469–1476.

Barton, H. A., Tang, J., Sey, S. M., Stanko, J. P., Murrell, R. N., Rockett, J. C., et al. (2006). Metabolism of myclobutanil and triadimefon by human and rat cytochrome P450 enzymes and liver microsomes. *Xenobiotica*, *36*, 793–806.

Bentley, E. W. (1969). The warfarin resistance problem in England and Wales. *Schriftenr. Ver. Wasser-Boden-Lufthyg.*, *32*, 19.

Beumel, G. A., Levi, P. E., & Hodgson, E. (1985). Spectral interactions of piperonyl butoxide and isocyanides with purified hepatic cytochrome P-450 from uninduced mice. *Gen. Pharmacol.*, *16*, 193–197.

Bhamre, S., Bhagwat, S., Shankar, S. K., Williams, D. E., & Ravindranath, V. (1993). Cerebral flavin-containing monooxygenase-mediated metabolism of antidepressants in brain: Immunochemical properties and immunocyto-chemical localization. *J. Pharmacol. Exp. Ther.*, *267*, 555–559.

Bikadi, Z., & Hazai, E. (2008). In silico description of differential enantioselectivity in methoxychlor O-demethylation by CYP2C enzymes. *Biochim. Biophys. Acta.*, *1780*, 1070–1079.

Blake, B. L., Philpot, R. M., Levi, P. E., & Hodgson, E. (1996). Xenobiotic biotransforming enzymes in the central nervous system: An isoform of flavin-containing monooxygenase FMO4 is expressed in rabbit brain. *Chem. Biol. Interact.*, *99*, 253–261.

Bogen, K. T., Benson, J. M., Yost, G. S., Morris, J. B., Dahl, A. R., Clewell, H. J., et al. (2008). Naphthalene metabolism in relation to target tissue anatomy, physiology, cytotoxicity and tumorigenic mechanism of action. *Regul. Toxicol. Pharmacol.*, *51*, S27–S36.

Boland, B., Lin, C.Y., Morin, D., Miller, L., Plopper, C., & Buckpitt, A. (2004). Site-specific metabolism of naphthalene and 1-nitronaphthalene in dissected airways of rhesus macaques. *J. Pharmacol. Exp. Ther.*, *310*, 546–554.

Bonate, P. L. (1991). Gender-related differences in xenobiotic metabolism. *J. Clin. Pharmacol.*, *31*, 684–690.

Bond, J. A. (1983). Some biotransformation enzymes responsible for polycyclic aromatic hydrocarbon metabolism in the rat nasal turbinates: Effects on enzyme activities of *in vitro* modifiers and intraperitoneal and inhalation exposure of rats to inducing agents. *Cancer Res.*, *43*, 4805–4811.

Bond, J. A. (1993). Metabolism and elimination of inhaled drugs and airborne chemicals from the lungs. *Pharmacol. Toxicol.*, *72*, 36–47.

Boyd, C. E., & Ferguson, D. E. (1964). Spectrum of cross-resistance to insecticides in the mosquito fish *Gambusia affinis*. *Mosq. News*, *24*, 19–21.

Boyd, C. E., & Ferguson, D. E. (1964). Susceptibility and resistance of mosquito fish to several insecticides. *J. Econ. Entomol.*, *57*, 430–431.

Boyd, C. E., Vinson, S. B., & Ferguson, D. E. (1963). Possible DDT resistance in two species of frogs. *Copeia*, *1963*, 426–429.

Boyle, C. M. (1960). Case of apparent resistance of *Rattus norvegicus* Berkenhout to anticoagulant poisons. *Nature*, *188*, 517.

Britto, M. R., & Wedlund, P. J. (1992). Cytochrome P-450 in the brain: Potential evolutionary and therapeutic relevance of localization of drug-metabolizing enzymes. *Drug Metab. Dispos.*, *20*, 446.

Brodie, B. B., & Maickel, R. P. (1962). Comparative biochemistry of drug metabolism. In *Proceedings of the 1st International Pharmacology Meeting*, 1961 (Vol. 6, pp. 299–324).

Brothers, D. R. (1972). A case of anticoagulant rodenticide resistance in an Idaho Norway rat *Rattus norvegicus* population. *Calif. Vector Views*, *19*, 41–45.

Brown, A. W. A., & Pal, R. (1971). *Insecticide resistance in arthropods* (2nd ed.). Geneva: World Health Organization.

Brown, M. A., Kimmel, E. C., & Casida, J. E. (1988). DNA adduct formation by alachlor metabolites. *Life Sci.*, *43*, 2087–2094.

Buratti, F. M., & Testai, E. (2005). Malathion detoxication by human carboxylesterases and its inhibition by isomalathion and other pesticides. *J. Biochem. Mol. Toxicol.*, *19*, 406–414.

Buratti, F. M., & Testai, E. (2007). Evidence for CYP3A4 autoactivation in the desulfuration of dimethoate by the human liver. *Toxicology*, *241*, 33–46.

Buratti, F. M., Volpe, M. T., Fabrizi, L., Meheguz, A., Vitozzi, L., & Testai, E. (2002). Kinetic parameters of OPT pesticide desulfuration by c-DNA expressed human CYPs. *Environ. Toxicol. Pharmacol.*, *11*, 181–190.

Buratti, F. M., Volpe, M. T., Meneguz, A., Vitossi, L., & Testai, E. (2003). CYP-specific bioactivation of four organophosphorothioate pesticides by human liver microsomes. *Toxicol. Appl. Pharmacol.*, *186*, 143–154.

Buratti, FM, D'Aniello, A, Volpe, MT, Meneguz, A, & Testai, E. (2005). Malathion bioactivation in the human liver: The contribution of different cytochrome P450 isoforms. *Drug Metab. Dispos.*, *33*, 295–302.

Buratti, FM, Leoni, C, & Testai, E. (2006). Foetal and adult CYP3A isoforms in the bioactivation of organophosphorothionate insecticides. *Toxicol. Lett.*, *167*, 245–255.

Buratti, F. M., Leoni, C., & Testai, E. (2007). The human metabolism of organophosphorothionate pesticides: Consequences for toxicological risk assessment. *J. Consumer Prot. Food Saf.*, *2*, 37–44.

Burke, M. D., & Mayer, R. T. (1974). Ethoxyresorufin: Direct fluorimetric assay of a microsomal O-dealkylation which is preferentially inducible by 3-methylcholanthrene. *Drug Metab. Dispos.*, *2*, 583–588.

Burnett, V. L., Lawton, M. P., & Philpot, R. M. (1994). Cloning and sequencing of flavin-containing monooxygenases FMO3 and FMO4 from rabbit and characterization of FMO3. *J. Biol. Chem.*, *269*, 14314–14322.

Buronfosse, T., Moroni, P., Benoit, E., & Riviere, J. L. (1995). Stereoselective sulfoxidation of the pesticide methiocarb by flavin-containing monooxygenase and cytochrome P450-dependent monooxygenases of rat liver microsomes: Anticholinesterase activity of the two sulfoxide enantiomers. *J. Biochem. Toxicol.*, *10*, 179–189.

Butler, A. M., & Murray, M. (1997). Biotransformation of parathion in human liver: Participation of CYP3A4 and its inactivation during parathion oxidation. *J. Pharmacol. Exp. Ther.*, *280*, 966–973.

Cain, D., Hutson, S., & Wallin, R. (1998). Warfarin resistance is associated with a protein component of the vitamin K 2,3-epoxide reductase enzyme complex in rat liver. *Thromb. Haemostasis*, *80*, 128–133.

Campbell, H. A., Smith, W. K., Roberts, W. L., & Link, K. P. (1941). Studies on hemorrhagic sweet clover disease. II. Bioassay of hemorrhagic concentrates by following prothrombin level in plasma of rabbit blood. *J. Biol. Chem.*, *138*, 1–20.

Carver, M. P., Levi, P. E., & Riviere, J. E. (1990). Parathion metabolism during percutaneous absorption in perfused porcine skin. *Pestic. Biochem. Physiol.*, *38*, 245–254.

Casabar, R. C. T., Wallace, A. D., Hodgson, E., & Rose, R. L. (2006). Metabolism of endosulfan-alpha by human liver microsomes and its utility as a simultaneous in vitro probe for CYP2B6 and CYP3A4. *Drug Metab. Dispos.*, *34*, 1779–1785.

Cashman, J. R., & Hanzlik, R. P. (1981). Microsomal oxidation of thiobenzamide: A photometric assay for the flavin-containing monooxygenase. *Biochem. Biophys. Res. Commun.*, *98*, 147.

Cashman, J. R., & Zhang, J. (2006). Human flavin-containing monooxygenases. *Annu. Rev. Pharmacol. Toxicol.*, *46*, 65–100.

Casida, J. E., Ueda, K., Gaughan, L. C., Jao, L. T., & Soderlund, D. M. (1975–1976). Structure–biodegradability relationships in pyrethroid insecticides. *Arch. Environ. Contam. Toxicol.*, *3*, 491–500.

Cerrara, G., & Periquet, A. (1991). Metabolism and toxicokinetics of pesticides in animals. In T. S. S. Dikshith (Ed.), *Toxicology of pesticides in animals* (pp. 67–118). Boca Raton, FL: CRC Press.

Chadwick, R. W., Chuang, L. T., & Williams, K. (1975). Dehydrogenation: A previously unreported pathway of lindane metabolism in mammals. *Pestic. Biochem. Physiol.*, *5*, 575–586.

Chambers, J. E., & Chambers, H. W. (1989). Oxidative desulfuration of chlorpyrifos, chlorpyrifos-methyl, and leptophos by rat brain and liver. *J. Biochem. Toxicol.*, *4*, 201–203.

Chambers, H. W., Browne, B., & Chambers, J. E. (1990). Noncatalytic detoxication of six organophosphorus compounds by rat liver homogenates. *Pestic. Biochem. Physiol.*, *36*, 308–315.

Chang, S. K., Williams, P. L., Dauterman, W. C., & Riviere, J. E. (1994). Percutaneous absorption, dermatopharmacokinetics and related bio-transformation studies of carbaryl, lindane, malathion, and parathion in isolated perfused porcine skin. *Toxicology*, *91*, 269–280.

Cherrington, N. J., Can, Y., Cherrington, J. W., Rose, R. L., & Hodgson, E. (1998). Physiological factors affecting protein expression of flavin-containing monooxygenases 1, 3 and 5. *Xenobiotica*, *7*, 673–682.

Cherrington, N. J., Falls, J. G., Rose, R. L., Clements, K. M., Philpot, R. M., & Levi, P. E., et al. (1998). Molecular cloning, sequence, and expression of mouse flavin-containing monooxygenases 1 and 5 FMO1 and FMO5. *J. Biochem. Mol. Toxicol.*, *12*, 205–212.

Chin, B. H., Eldridge, J. M., Anderson, J. H., & Sullivan, L. J. (1979). Carbaryl metabolism in the rat: A comparison of *in vivo, in vitro* tissue explant and liver perfusion techniques. *J. Agric. Food Chem.*, *27*, 716–720.

Chin, B. H., Sullivan, L. J., & Eldridge, J. M. (1979). *In vitro* metabolism of carbaryl by liver explants of bluegill, catfish, perch, goldfish and kissing gourami. *J. Agric. Food Chem.*, *27*, 1395–1398.

Chin, B. H., Sullivan, L. J., Eldridge, J. M., & Tallant, M. J. (1979). Metabolism of carbaryl by kidney, liver, and lung from human postembryonic fetal autopsy tissue. *Clin. Toxicol.*, *14*, 489–498.

Cho, H. Y., & Kong, K. H. (2007). Study on the biochemical characterization of herbicide detoxification enzyme, glutathione S-transferase. *Biofactors*, *30*, 281–287.

Cho, T. M., Rose, R. L., & Hodgson, E. (2006). In vitro metabolism of naphthalene by human liver microsomal cytochrome P450 enzymes. *Drug Metab. Dispos.*, *34*, 176–183.

Choi, J., Rose, R. L., & Hodgson, E. (2002). In vitro human metabolism of permethrin: The role of alcohol and aldehyde dehydrogenases. *Pestic. Biochem. Physiol.*, *73*, 117–128.

Choi, K., Joo, H., Rose, R. L., & Hodgson, E. (2006). Metabolism of chlorpyrifos and chlorpyrifos oxon by human hepatocytes. *J. Biochem. Mol. Toxicol.*, *20*, 279–291.

Class, T. J., Ando, T., & Casida, J. E. (1991). Pyrethroid metabolism: Microsomal oxidase metabolites of S-bioallethrin and the six natural pyrethrins. *J. Agric. Food Chem.*, *38*, 529–537.

Coleman, S., Liu, S., Linderman, R., Hodgson, E., & Rose, R. L. (1999). *In vitro* metabolism of alachlor by human liver microsomes and human cytochrome P450 isoforms. *Chem. Biol. Interact.*, *122*, 27–39.

Coleman, S., Linderman, R., Hodgson, E., & Rose, R. L. (2000). Comparative metabolism of chloracetamide herbicides and selected metabolites in human and rat liver microsomes. *Environ. Health Perspect.*, *108*, 1151–1157.

Conney, A. H., & Burns, J. J. (1962). Factors influencing drug metabolism. *Adv. Pharmacol.*, *1*, 31–58.

Cortright, K. A., & Craigmill, A. L. (2006). Cytochrome P450-dependent metabolism of midazolam in hepatic microsomes from chickens, turkeys, pheasant and bobwhite quail. *J. Vet. Pharmacol. Ther.*, *29*, 469–476.

Coutts, R. T., & Urichuk, L. J. (1999). Polymorphic cytochromes P450 and drugs used in psychiatry. *Cell. Mol. Neurol.*, *19*, 325–354.

Cram, R. L., Juchau, M. R., & Fouts, J. R. (1965). Differences in hepatic drug metabolism in various rabbit strains before and after pretreatment with phenobarbital. *Proc. Soc. Exp. Biol. Med.*, *118*, 872–875.

Croom, E. L., Stevens, J. C., Hines Ronald, N., Wallace, A. D., & Hodgson, E. (2009). Human hepatic CYP2B6 developmental expression: The impact of age and genotype. *Biochem. Pharmacol.*, *78*, 184–190.

Croom, E. L., Wallace, A. D., & Hodgson, E. (2010). Human variation in CYP-specific chlorpyrifos metabolism. *Toxicology*, *276*, 184–191.

Crow, J. A., Borazjani, A., Potter, P. M., & Ross, M. K. (2007). Hydrolysis of pyrethroids by human and rat tissues: Examination of intestinal, liver and serum carboxylesterases. *Toxicol. Appl. Pharmacol.*, *221*, 1–12.

Dahl, A. R., & Hodgson, E. (1979). The interaction of aliphatic analogs of methylene-dioxyphenyl compounds with cytochromes P450 and P430. *Chem. Biol. Interact.*, *27*, 163–175.

Dahl, A. R., & Lewis, J. L. (1993). Respiratory tract uptake of inhalants and metabolism of xenobiotics. *Annu. Rev. Pharmacol. Toxicol., 32*, 383–407.

Dai, D., Tang, J., Rose, R., Hodgson, E., Bienstock, R. J., Mohrenweiser, H. W., et al. (2001). Identification of variants of CYP3A4 and characterization of their abilities to metabolize testosterone and chlorpyrifos. *J. Pharmacol. Exp. Ther., 299*, 825–831.

Daly, A. K., Fairbrother, K. S., & Smart, J. (1998). Recent advances in understanding the molecular basis of polymorphisms in genes encoding cytochrome P450 enzymes. *Toxicol. Lett., 102*, 143–147.

Dannan, G. A., Guengerich, F. P., & Waxman, D. J. (1986). Hormonal regulation of rat liver microsomal enzymes: Role of gonadal steroids in programming, maintenance, and suppression of δ-4-steroid 5a-reductase, flavin-containing monooxygenase, and sex-specific cytochromes P-450. *J. Biol. Chem., 261*, 10728–10735.

Dauterman, W. C. (1994). Adaptation to toxicants. In E. Hodgson & P. E. Levi (Eds.), *Introduction to biochemical toxicology* (pp. 569–581) (2nd ed.). Norwalk, CT: Appleton & Lange.

Deamer, N. J., O'Callaghan, J. P., & Genter, M. B. (1994). Olfactory toxicity resulting from dermal application of 2,6-dichlorobenzonitrile dichlobenil in the C57B1 mouse. *Neurotoxicology, 15*, 287–294.

Dehal, S. S., & Kupfer, D. (1994). Metabolism of the proestrogenic pesticide methoxychlor by hepatic P450 monooxygenases in rats and humans: Dual pathways involving novel ortho ring-hydroxylation by CYP2B. *Drug Metab. Dispos., 22*, 937–946.

Denison, M. S., Murray, M., & Wilkinson, C. F. (1983). Microsomal aryl hydrocarbon hydroxylase: Comparison of the direct, indirect and radiometric assays. *Anal. Lett., 16B5*, 381–391.

Dick, R. A., Kanne, D. B., & Casida, J. E. (2005). Identification of aldehyde oxidase as the neonicotinoid reductase. *Chem. Res. Toxicol., 18*, 317–323.

Ding, X., & Kamienski, L. S. (2003). Human extrahepatic cytochromes P450: Function in xenobiotic metabolism and tissue-selective chemical toxicity in the respiratory and gastrointestinal tracts. *Annu. Rev. Pharmacol. Toxicol., 43*, 149–173.

Ding, X., Sheng, J., & Bhama, J. K. (1994). Metabolic activation of a potent olfactory-specific toxicant, 2,6-dichlorobenzonitrile CNB by P450 2As. Paper 309, 16th Annual Meeting of the Association for Chemoreception Sciences, Sarasota, FL, 1994.

Ding, X., Spink, D. C., Bhama, J. K., Sheng, J. J., Vaz, A. D., & Coon, M. J. (1996). Metabolic activation of 2,4-dichlorobenzonitrile, an olfactory-specific toxicant, by rat, rabbit, and human cytochromes P450. *Mol. Pharmacol., 49*, 1113–1121.

Dixit, A., & Roche, T. E. (1984). Spectrophotometric assay of the flavin-containing monooxygenase and changes in its activity in female mouse liver with nutritional and diurnal conditions. *Arch. Biochem. Biophys., 233*, 50–60.

Donninger, C., Hutson, D. H., & Pickering, B. A. (1967). Oxidative cleavage of phosphoric acid triesters to diesters. *Biochem. J., 102*, 26.

Donninger, C., Hutson, D. H., & Pickering, B. A. (1972). The oxidative dealkylation of insecticidal phosphoric acid triesters by mammalian liver enzymes. *Biochem. J., 126*, 701–707.

Dorough, H. W. (1968). Metabolism of furadan. *J. Agric. Food Chem., 16*, 319–324.

Dorough, H. W. (1970). Metabolism of insecticidal methylcarbamates in animals. *J. Agric. Food Chem., 18*, 1015–1022.

Dorough, H. W. (1984). Metabolism of insecticides by conjugation mechanisms. In F. Matsumura (Ed.), *International encyclopedia of pharmacology and therapeutics.* Section 113. *Differential toxicities of insecticides and halogenated aromatics.* New York: Pergamon Press.

Dorough, H. W., & Casida, J. E. (1964). Nature of certain carbamate metabolites of the insecticide Sevin. *J. Agric. Food Chem., 12*, 294–304.

Douch, P. G. C., & Smith, J. N. (1971). Metabolism of *m*-tert-butylphenyl, *N*-methylcarbamate in insects and mice. *Biochem. J., 125*, 385–393.

Douch, P. G. C., & Smith, J. N. (1971). The metabolism of 3,5-di-tert-butylphenyl, *N*-methylcarbamate in insects and by mouse liver enzymes. *Biochem. J., 125*, 395–400.

Drummond, D. C. (1966). Rats' resistance to warfarin. *New Sci., 30*, 771–772.

Duffel, M. W., & Gillespie, S. G. (1984). Microsomal flavin-containing monooxygenase activity in rat corpus striatum. *J. Neurochem., 42*, 1350–1353.

Duffel, M. W., Graham, J. M., & Ziegler, D. M. (1981). Changes in dimethylaniline N-oxidase activity of mouse liver and kidney induced by steroid sex hormones. *Mol. Pharmacol., 19*, 134–139.

Dupuy, J., Escudero, E., Eeckhoutte, C., Sutra, J. F., Galtier, P., & Alvinerie, M. (2001). In vitro metabolism of C14-moxidectin by hepatic microsomes from various species. *Vet. Res., 25*, 345–354.

Ecobichon, D. J. (2001). Toxic effects of pesticides. In C. D. Klaassen (Ed.), *Casarett and doull's toxicology.* New York: McGraw–Hill.

Esaac, E. G., & Matsumura, F. (1984). Metabolism of insecticides by reductive systems. In F. Matsumura (Ed.), *International encyclopedia of pharmacology and therapeutics.* Section 113. *Differential toxicities of insecticides and halogenated aromatics* (pp. 265–290). New York: Pergamon.

Eto, M., Casida, J. E., & Eto, T. (1962). Hydroxylation and cyclization reactions involved in the metabolism of tri-*o*-cresyl phosphate. *Biochem. Pharmacol., 11*, 337–352.

Evered, D., & Collins, G. M. (1984). Origins and development of adaptation. *Ciba Found. Symp., 102*, 273.

Falls, J. G., Cao, Y., Blake, B. L., Levi, P. E., & Hodgson, E. (1995). Gender differences in hepatic expression of flavin-containing monooxygenase isoforms FMO1, FMO3, and FMO5 in mice. *J. Biochem. Toxicol., 10*, 171–177.

Falls, J. G., Ryu, D.-Y., Cao, Y., Levi, P. E., & Hodgson, E. (1997). Regulation of mouse liver flavin-containing monooxygenases 1 and 3 by sex steroids. *Arch. Biochem. Biophys., 342*, 212–223.

Feng, P. C. C., & Wratten, S. J. (1989). In vitro transformation of chloroacetanilide herbicides by rat liver enzymes: A comparative study of metolachlor and alachlor. *J. Agric. Food Chem., 37*, 1088–1093.

Feng, P. C. C., Wilson, A. G. E., McClanahan, R. H., Patanella, J. E., & Wratten, S. J. (1990). Metabolism of alachlor by rat and mouse liver and nasal turbinate tissues. *Drug Metab. Dispos., 18*, 373–377.

Feng, P. C., Solsjen, R. T., & McClanahan, R. H. (1994). In vitro transformation of thiozopyr by rat liver enzymes: Sulfur and carbon oxidations by microsomes. *Pestic. Biochem. Biophys., 48*, 8.

Ferguson, D. E., & Bingham, C. R. (1966). The effects of combinations of insecticides on susceptible and resistant mosquito fish. *Bull. Environ. Contam. Toxicol., 1*, 97–103.

Ferguson, D. E., & Bingham, C. R. (1966). Endrin resistance in yellow bullheads, *Ictalurus natalis. Trans. Am. Fish. Soc., 95*, 325–326.

Ferguson, D. E., & Boyd, C. E. (1964). Apparent resistance to methyl parathion in mosquito fish, *Gambusia affinis. Copeia, 4*, 706.

Ferguson, D. E., Culley, D. D., Cotton, W. D., & Dodds, R. P. (1964). Resistance to chlorinated hydrocarbon insecticides in three species of fresh water fish. *BioScience, 14*, 43–44.

Ferguson, D. E., Cotton, W. D., Gardner, D. T., & Culley, D. D. (1965). Tolerances to five chlorinated hydrocarbon insecticides in two species of fish from a transect of the lower Mississippi River. *J. Miss. Acad. Sci., 11*, 239–245.

Figueiredo-Fernandes, A., Fontainhas-Fernandes, A., Rocha, E., & Reis-Henriques, M. A. (2006). The effect of paraquat on hepatic EROD activity and gonadal histology in males and females of Nile tilapia, Oreochromis niloticus, exposed at different temperatures. *Arch. Environ. Contam. Toxicol., 51*, 626–632.

Fouts, J. R., & Adamson, R. H. (1959). Drug metabolism in the new born rabbit. *Science, 129*, 897–898.

Foxenberg, R. J., McGarrigle, B. P., Knaak, J. B., Kostyniak, P. J., & Olson, J. R. (2007). Human hepatic cytochrome P45-specific metabolism of parathion and chlorpyrifos. *Drug Metab. Dispos., 35*, 189–193.

Fredriksson, T., Farrior, W. L., Jr., & Witter, R. F. (1961). Studies on the percutaneous absorption of parathion and paraoxon. I. Hydrolysis and metabolism within the skin. *Acta Derm. Venereol., 41*, 335–343.

Fujioka, K, & Casida, J. E. (2007). Glutathione S-transferase conjugation of organophosphorus pesticides yields S-phospho, S-aryl and S-alkylglutathione derivatives. *Chem. Res. Toxicol., 20*, 1211–1217.

Furnes, B, & Schlenk, D. (2004). Evaluation of xenobiotic N- and S-oxidation by variant flavin-containing monooxygenase 1 (FMO1) enzymes. *Toxicol. Sci., 78*, 196–203.

Furnes, B., & Schlenk, D. (2005). Extrahepatic metabolism of carbamate and organophosphate thioether compounds by the flavin-containing monooxygenase and cytochrome P450 systems. *Drug Metab. Dispos., 33*, 214–218.

Gelboin, H. V., & Conney, A. H. (1968). Antagonism and potentiation of drug action. In E. Boyland & R. Goulding (Eds.), *Modern trends in toxicology* (pp. 175–195). London: Butterworths.

Genter, M. B., Owens, D. M., & Deamer, N. J. (1995). Distribution of microsomal epoxide hydrolase and glutathione S-transferase enzymes in the rat olfactory mucosa: Relevance to distribution of lesions caused by systemically administered factory toxicants. *Chem. Senses, 20*, 385–392.

Genter, M. B., Deamer-Melia, N. J., Wetmore, B. A., Morgan, K. T., & Meyer, S. A. (1998). Herbicides and olfactory/neurotoxicity responses. *Rev. Toxicol.*, *2*, 93–112.

Georghiou, G. P., & Saito, T. (1983). *Pest resistance to pesticides*. New York: Plenum.

Ghersi-Egea, J. F., Perrin, R., Leininger-Miller, B., Grassiot, M. C., Jeandel, C., Floquet, J., et al. (1993). Subcellular localization of cytochrome P450, and activities of several enzymes responsible for drug metabolism in the human brain. *Biochem. Pharmacol.*, *45*, 647.

Ghiasuddin, S. M., & Soderlund, D. M. (1984). Hydrolysis of pyrethroid insecticides by soluble mouse brain esterases. *Toxicol. Appl. Pharmacol.*, *74*, 390–396.

Glusczak, L., Miron, D. D., Moraes, B. S., Simoes, R. R., Schelinger, M. R. C., Morsch, V. M., et al. (2007). Acute effects of glyphosate herbicide on metabolic and enzymatic parameters of silver catfish Rhamdia quelen. *Comp. Biochem. Physiol. C Toxicol. Pharmacol.*, *146*, 519–524.

Godin, S. J., Scollon, E. J., Hughes, M. F., Potter, P. M., DeVito, M. J., & Ross, M. K. (2006). Species differences in the in vitro metabolism of deltamethrin and esfenvalerate: Differential oxidative and hydrolytic metabolism by humans and rats. *Drug Metab. Dispos.*, *34*, 1764–1771.

Godin, S. J., Crow, J. A., Scollon, E. J., Hughes, M. F., DeVito, M. J., & Ross, M. K. (2007). Identification of rat and human cytochrome P450 isoforms and a rat serum esterase that metabolize the pyrethroid insecticides deltamethrin and esfenvalerate. *Drug Metab. Dispos.*, *35*, 1664–1671.

Goerz, G., Bolsen, K., Kalofoutis, A., & Tsambaos, D. (1994). Influence of oral isotretinoin on hepatic and cutaneous P-450-dependent isozyme activities. *Arch. Dermatol. Res.*, *286*, 104–106.

Goldstein, J. A., & De Morais, S. M. F. (1994). Biochemistry and molecular biology of the human CYP2C subfamily. *Pharmacogenetics*, *4*, 285–299.

Gonzalez, F. J. (1989). The molecular biology of cytochrome P-450s. *Pharmacol. Rev.*, *40*, 243–287.

Gonzalez, F. J. (1990). Molecular genetics of the P450 superfamily. *Pharmacol. Ther.*, *45*, 1–38.

Gonzalez, F. J. (2003). Role of gene knockout and transgenic mice in the study of xenobiotic metabolism. *Drug Metab. Rev.*, *35*, 319–335.

Guengerich, F. P. (1992). Metabolic activation of carcinogens. *Pharmacol. Ther.*, *54*, 17–61.

Guengerich, F. P. (1993). Bioactivation and detoxication of toxic and carcinogenic chemical. *Drug Metab. Dispos.*, *21*, 1–6.

Gunderson, M. P., Kohno, S., Blumberg, B., Iguchi, T., & Guillette, L. J. (2006). Up-regulation of the alligator CYP3A77 gene by toxaphene and dexamethasone and its short term effect on plasma testosterone concentrations. *Aquat. Toxicol.*, *78*, 272–283.

Guthrie, F. E., Monroe, R. J., & Abernathy, C. O. (1971). Response of the laboratory mouse to selection for resistance to insecticides. *Toxicol. Appl. Pharmacol.*, *18*, 92–101.

Hackett, A. G., Kotyr, J. J., & Fujiwara, H., et al. (1993). Metabolism of triallate in Sprague-Dawley rats. 3. In vitro metabolic pathways. *J. Agric. Food Chem.*, *41*, 141.

Hainzl, D., & Casida, J. E. (1996). Fipronil insecticide: Novel photochemical desulfinylation with retention of neurotoxicity. *Proc. Natl. Acad. Sci. USA*, *93*, 12764–12767.

Hajjar, N. P., & Hodgson, E. (1980). Flavin adenine dinucleotide-dependent monooxygenase: Its role in the sulfoxidation of pesticides in mammals. *Science*, *209*, 1134–1136.

Hajjar, N. P., & Hodgson, E. (1982). The microsomal FAD-dependent monooxygenase as an activating enzyme: Fonofos metabolism. In R. Synder, D. V. Parke, J. J. Kocsis, D. Jollow, G. G. Gibson, & C. Witmer (Eds.), *Biological reactive intermediates* (Vol. 2, pp. 1245–1253). New York: Plenum. Part B.

Hajjar, N. P., & Hodgson, E. (1982). Sulfoxidation of thioether-containing pesticides by the flavin-adenine dinucleotide-dependent monooxygenase of pig liver microsomes. *Biochem. Pharmacol.*, *31*, 745–752.

Hanioka, N., Jinno, H., Kitazawa, K., Tanaka-Kagawa, T., Nishimura, T., Ando, M., et al. (1998). In vitro biotransformation of atrazine by rat liver microsomal cytochrome P450 enzymes. *Chem. Biol. Interact.*, *116*, 181–198.

Hanioka, N., Jinno, H., Tanaka-Kagawa, T., Nishimura, T., & Ando, M. (1999). In vitro metabolism of simazime, atrazine and propazine by hepatic cytochrome P450 enzymes of rat mouse and guinea pig, and oestrogenic activity of chlorotriazines and their main metabolites. *Xenobiotica.*, *29*, 1213–1226.

Hansson, T., von Bahr, C., Marklund, M., Svensson, J. O., Ingelman-Sundberg, M., & Lundstrom, J. (1992). Different regiospecificity in the hydroxylation of the antidepressant desmethylimipramine between rat brain and liver. *Pharmacol. Toxicol.*, *71*, 416.

Hayes, J. D., Pickett, C. B., & Mantle, T. J. (Eds.). (1990). *Glutathione S-transferase and drug resistance*. London: Taylor & Francis.

Hazai, E., Gagne, P. V., & Kupfer, D. (2004). Glucuronidation of the oxidative cytochrome P450-mediated phenolic metabolites of the endocrine disruptor pesticide methoxychlor by human hepatic UDP-glucuronosyl transferases. *Drug Metab. Dispos.*, *32*, 742–751.

Henderson, M. C., Krueger, S. K., Siddens, L. K., Stevens, J. F., & Williams, D. E., (2004). S-oxygenation of the thioether organophate insecticides phorate and disulfoton by human lung flavin-containing monooxygenase, 2. *Biochem. Pharmacol.*, *68*, 959–967.

Heymann, E., & Mentlein, R. (1981). Carboxylesterases—amidases. In W. B. Jacoby (Ed.), *Methods in enzymology* (Vol. 77, pp. 333–344). New York: Academic Press.

Hitchcock, M., & Murphy, S. D. (1967). Enzymatic reduction of *O,O*-diethyl-4-nitrophenyl phosphorothioate, *O,O*-diethyl *O*-4-nitrophenyl phosphate, and *o*-ethyl *O*-4-nitrophenyl benzene thiophosphonate by tissues from mammals, birds and fishes. *Biochem. Pharmacol.*, *16*, 1801–1811.

Hodgson, A.V., White, T. B., White, J. W., & Strobel, H. W. (1993). Expression analysis of the mixed function oxidase system in rat brain by the polymerase chain reaction. *Mol. Cell. Biochem.*, *120*, 171–179.

Hodgson, E. (1979). Comparative aspects of the distribution of the cytochrome P450-dependent monooxygenase system: An overview. *Drug Metab. Rev.*, *10*, 15–33.

Hodgson, E. (2003). In vitro human phase I metabolism of xenobiotics. I. Pesticides and related compounds used in agriculture and public health, May 2003. *J. Biochem. Mol. Toxicol.*, *17*, 201–206.

Hodgson, E. (2011). In vitro metabolism of agrochemicals and related chemicals. Foundation for Toxicology and Agromedicine <www.toxicologyagromed.org> .

Hodgson, E., & Casida, J. E. (1961). Metabolism of *N,N*-dialkyl carbamates and related compounds by rat liver. *Biochem. Pharmacol.*, *8*, 179–191.

Hodgson, E., & Casida, J. E. (1962). Mammalian enzymes involved in the degradation of *O,O*-dimethyl 2,2-dichlorovinyl phosphate Vapona or DDVP. *J. Agric. Food Chem.*, *10*, 208–214.

Hodgson, E., & Croom, E. L. (2008). Phase I—toxicogenetics. In R. C. Smart & E. Hodgson (Eds.), *Molecular and biochemical toxicology*. Hoboken, NJ: John Wiley & Sons. (Chapter 11)

Hodgson, E., & Levi, P. E. (1992). The role of the flavin-containing monooxygenase EC 1.14.13.8 in the metabolism and mode of action of agricultural chemicals. *Xenobiotica*, *22*, 1175–1183.

Hodgson, E., & Levi, P. E. (2001). Metabolism of pesticides (2nd ed.). In R. I. (2001). Krieger (Ed.), *Handbook of pesticide toxicology* (Vol. 1, pp. 531–562). San Diego: Academic Press.

Hodgson, E., & Meyer, S. A. (1997). Pesticides. In I. G. Sipes, C. A. McQueen, & A. J. Gandolfi (Eds.), *Comprehensive toxicology*. Vol. 9. *Hepatic and gastrointestinal toxicology*. In R. S. McCuskey, & D. L. Earnest (Eds.), New York: Elsevier.

Hodgson, E., & Meyer, S. A. (2010). Pesticides and hepatotoxicity. In C. A. McQueen (Ed.), *Comprehensive Toxicology* (2nd ed., Vol. 9). In P. E. Ganey, & R. Roth (Eds.), *Hepatic toxicology*. New York: Elsevier.

Hodgson, E., & Rose, R. L. (2005). Human metabolism and metabolic interactions of deployment-related chemicals. *Drug Metab. Rev.*, *37*, 1–39.

Hodgson, E., & Rose, R. L. (2007). The importance of P450 2B6 CYP2B6 in the human metabolism of environmental chemicals. *Pharmacol. Ther.*, *113*, 420–428.

Hodgson, E., Silver, I. S., Butler, L. E., Lawton, M. P., & Levi, P. E. (1991). In W. J. Hayes, Jr. & E. R. Laws Jr. (Eds.), *Handbook of pesticide toxicology* (Vol. 1, pp. 107–167). San Diego: Academic Press.

Hodgson, E., Cherrington, N., Coleman, S. C., Liu, S., Falls, J. G., Cao, Y., et al. (1998). Flavin-containing monooxygenase and cytochrome P450 mediated metabolism of pesticides: From mouse to human. *Rev. Toxicol.*, *2*, 231–243.

Hollingworth, R. M. (1969). Dearylation of organophosphorus esters by mouse liver enzymes in vitro and in vivo. *J. Agric. Food Chem.*, *17*, 987–996.

Hollingworth, R. M., Kurihara, N., Miyamoto, J., Otto, S., & Paulson, G. D. (1995). Detection and significance of active metabolites of agrochemicals and related xenobiotics in animals. *Pure Appl. Chem.*, *67*, 1487–1532.

Honda, H., Tomizawa, M., & Casida, J. E. (2006). Neonicotinoid metabolic activation and inactivation established with coupled nicotin receptor–CYP3A4 and –aldehyde oxidase systems. *Toxicol. Lett*, *161*, 108–114.

Hosteter, K. A., Wrighton, S. A., Kremers, P., & Guzelian, P. S. (1987). Immunological evidence for multiple steroid-inducible hepatic cytochromes P-450 in rat. *Biochem. J.*, *245*, 27–33.

Hu, J. J., Lee, M. J., Vapiwala, M., Reuhl, K., Thomas, P. E., & Yang, C. S. (1993). Sex-related differences in mouse and renal metabolism and toxicity of acetaminophen. *Toxicol. Appl. Pharmacol.*, *122*, 16–26.

Hu, Y. D., & Kupfer, D. (2002). Enantioselective metabolism of the endocrine disruptor pesticide methoxychlor by human cytochrome P450s: Major differences in selective enantiomer formation by various P450 isoforms. *Drug Metab. Dispos.*, *30*, 1329–1336.

Hu, Y. D., & Kupfer, D. (2002). Metabolism of the endocrine disruptor pesticide methoxychlor by human cytochrome P450s: Pathways involving a novel catechol metabolite. *Drug Metab. Dispos.*, *30*, 1035–1042.

Huang, H. Z., Fleming, C. D., Nishi, K., Redindo, M. R., & Hammock, B. D. (2005). Stereoselective hydrolysis of pyrethroid-like substrates by human and other mammalian carboxylesterases. *Chem. Res. Toxicol.*, *18*, 1371–1377.

Hucker, H. B., Gillette, J. R., & Brodie, B. B. (1960). Enzymic pathway for the formation of cotinine, a major metabolite of nicotine in rabbit liver. *J. Pharmacol. Exp. Ther.*, *129*, 94–100.

Hur, J. H., Wu, S. Y., & Casida, J. E. (1992). Oxidative chemistry and toxicology of S,S,S-tributyl phosphorothioate. *J. Agric. Food Chem.*, *40*, 1703–1709.

Hutson, D. H. (1976). Comparative metabolism of dieldrin in the rat CFE and two strains of mouse CFI and LACG. *Food Cosmet. Toxicol.*, *14*, 577–591.

Hutson, D. H. (1981). The metabolism of insecticides in man. *Prog. Pestic. Biochem.*, *1*, 287–333.

Jackson, W. B., & Kaukeinen, D. (1972). Resistance of wild Norway rats in North Carolina to warfarin rodenticide. *Science*, *176*, 1343–1344.

Jacobsen, N. E., Sanders, M., Toia, R. F., et al. (1991). Alachlor and its analogues as metabolic progenitors of formaldehyde: Fate of N-methoxymethyl and other N-alkoxyalkyl substituents. *J. Agric. Food Chem.*, *39*, 1342–1350.

James, M. O., Stuchal, L. D., & Nyagode, B. A. (2008). Glucuronidation and sulfonation, in vitro, of the major endocrine-active metabolites of methoxychlor in the channel catfish, Ictalurus punctatus, and induction following treatment with 3-methylcholanthrene. *Aquat. Toxicol.*, *86*, 227–238.

Jay, G. E., Jr. (1955). Variation in response of various mouse strains to hexobarbital Evipal. *Proc. Soc. Exp. Biol. Med.*, *90*, 378–380.

Jerina, D. M., Daly, J. W., Witkop, B., Zaltman-Nirenberg, P., & Uden-friend, S. (1968). The role of arene oxide-oxepin systems in the metabolism of aromatic substrates. III. Formation of 1,2-napthalene oxide from napthalene by liver microsomes. *J. Am. Chem. Soc.*, *90*, 6525–6527.

Jerina, D. M., Daly, J. W., Witkop, B., Zaltman-Nirenberg, P., & Udenfriend, S. (1970). 1,2-Naphthalene oxide as an intermediate in the microsomal hydroxylation of napthalene. *Biochemistry*, *9*, 147–155.

Joo, H, Choi, K, & Hodgson, E. (2010). Human metabolism of atrazine. *Pestic. Biochem. Physiol.*, *98*, 73–79.

Judah, J. D. (1949). Studies on the metabolism and mode of action of DDT. *Br. J. Pharmacol. Chemother*, *4*, 120–131.

Jugert, F. K., Agarwal, R., Kuhn, A., Bickers, D. R., Merk, H., & Muktar, H. (1994). Multiple cytochrome P450 isozymes in murine skin: Induction of P450 1A, 2B, 2E and 3A by dexamethasone. *J. Invest. Dermatol.*, *102*, 970–975.

Kalow, W. (1991). Interethenic variation of drug metabolism. *Trends Pharmacol. Sci.*, *12*, 102–107.

Kamataki, T., & Neal, R. A. (1976). Metabolism of diethyl-p-nitrophenyl phosphorothionate parathion by a reconstituted mixed function oxidase enzyme system: Studies of its covalent binding of the sulfur atom. *Mol. Pharmacol.*, *12*, 933–944.

Kapoor, I. P., Metcalf, R. L., & Nyström, R. F. (1970). Comparative metabolism of methoxychlor, methiochlor and DDT in mouse, insects and in a model ecosystem. *J. Agric. Food Chem.*, *20*, 1–16.

Kappers, W. A., Edwards, R. J., Murray, S., & Boobis, A. R. (2001). Diazinon is activated by CYP2C19 in human liver. *Toxicol. Appl. Pharmacol.*, *177*, 68–78.

Kawaji, A. K., Ohara, K., & Takabatake, E. (1994). Determination of flavin-containing monooxygenase activity in rat brain microsomes with benzydamine-N-oxidation. *Biol. Pharm. Bull.*, *17*, 603–606.

Kim, D. O., Lee, S. J., Jeon, T. W., Jin, C. H., Hyun, S. H., & Kim, E. J., et al. (2005). Role of metabolism in parathion-induced hepatotoxicity and immunotoxicity. *J. Toxicol. Environ. Health A Curr. Issues*, *66*, 2187–2205.

Kim, J. H., Lam, W. W., Quistad, G. B., & Cassida, J. E. (1994). Sulfoxidation of the soil fumigants metam, methyl isocyanate and dazomet. *J. Agric. Food Chem.*, *42*, 2019–2024.

Kimmel, E. C., Casida, J. E., & Ruzo, L. O. (1986). Formamidine insecticides and chloracetanilide herbicides: Disubstituted anilines and nitrosobenzenes as mammalian metabolites and bacterial mutagens. *J. Agric. Food Chem.*, *34*, 157–161.

Kinsler, S., Levi, P. E., & Hodgson, E. (1988). Hepatic and extrahepatic microsomal oxidation of phorate by the cytochrome P450 and FAD-containing monooxygenase systems in the mouse. *Pestic. Biochem. Physiol.*, *31*, 54–60.

Kinsler, S., Levi, P. E., & Hodgson, E. (1990). Relative contributions of the cytochrome P-450 and flavin-containing monooxygenases to the microsomal oxidation of phorate following treatment of mice with phenobarbital hydrocortisone, acetone and piperonyl butoxide. *Pestic. Biochem. Physiol.*, *37*, 174–181.

Kishimoto, D., Oku, A., & Kurihara, N. (1995). Enantiotropic selectivity of cytochrome P450-catalyzed oxidative demethylation of methoxychlor: Alteration of selectivity depending on isozymes and substrate concentration. *J. Agric. Food Chem.*, *51*, 12–19.

Kitahara, A., Ebina, T., Ishikawa, T., Soma, Y., Sata, K., & Kanai, S. (1982). Changes in activities and molecular forms of rat hepatic drug-metabolizing enzymes during aging. In K. Kitani (Ed.), *Liver and aging—liver and drugs* (pp. 135–142). Amsterdam: Elsevier Biomedical.

Kobliakov, V., Popova, N., & Rossi, L. (1991). Regulation of the expression of sex-specific isoforms of cytochrome P450 in rat liver. *Eur. J. Biochem.*, *195*, 588–591.

Koop, D. R. (1986). Hydroxylation of *p*-nitrophenol by rabbit ethanol inducible cytochrome P450 3A. *Mol. Pharmacol.*, *29*, 399–404.

Krieger, R. (Ed.). (2010). *Handbook of pesticide toxicology* (3rd edition). Amsterdam: Elsevier.

Krueger, S. K., & Williams, D. E. (2005). Mammalian flavin-containing monooxygenases: Structure/function, genetic polymorphisms, and role in drug metabolism. *Pharmacol. Ther.*, *106*, 357–387.

Kulkarni, A. P., & Hodgson, E. (1980). Metabolism of insecticides by mixed function oxidase systems. *Pharmacol. Ther.*, *8*, 379–475.

Kulkarni, A. P., & Hodgson, E. (1984a). Metabolism of insecticides by mixed function oxidase systems. In F. Matsumura (Ed.), *International encyclopedia of pharmacology and therapeutics*. Section 113. *Differential toxicities of insecticides and halogenated aromatics*. New York: Pergamon Press.

Kulkarni, A. P., & Hodgson, E. (1984b). The metabolism of insecticides: The role of monooxygenase systems. *Annu. Rev. Pharmacol. Toxicol.*, *24*, 19–42.

Kulkarni, A. P., Mailman, R. B., & Hodgson, E. (1975). Cytochrome P-450 optical difference spectra of insecticides: A comparative study. *J. Agric. Food Chem.*, *23*, 177–183.

Kulkarni, A. P., Levi, P. E., & Hodgson, E. (1984). The metabolism of insecticides: The role of monooxygenase enzymes. *Annu. Rev. Pharmacol. Toxicol.*, *24*, 19–42.

Kurihari, N., & Oka, A. (1991). Effects of added protein bovine serum albumin on the rate and enantiotropic selectivity of oxidative O-demethylation of methoxychlor in rat liver microsomes. *Pestic. Biochem. Physiol.*, *40*, 227–235.

Lai, K., Stolowich, N. J., & Wild, J. R. (1995). Characterization of P–S bond hydrolysis in organophosphorothioate pesticides by organophosphorus hydrolase. *Arch. Biochem. Biophys.*, *318*, 59–64.

Lang, D. H., Criegee, D., Grothusen, A., Saalfrank, R. W., & Bocker, R. H. (1996). In vitro metabolism of atrazine, terbutylazine, ametryne and terbutryne in rats, pigs and humans. *Drug Metab. Dispos.*, *24*, 859–865.

Lang, D. H., Rettie, A. E., & Bocker, R. H. (1997). Identification of enzymes involved in the metabolism of atrazine, ametryne and terbutryne in human liver microsomes. *Chem. Res. Toxicol.*, *10*, 1037–1044.

Lasseur, R., Longin-Sauvageon, C., Videmann, B., Billeret, M., Berny, P., & Benoit, E. (2005). Warfarin resistance in a French strain of rats. *J. Biochem. Molec. Toxicol.*, *19*, 379–385.

Lawton, M. P., Gasser, R., Tynes, R. E., Hodgson, E., & Philpot, R. M. (1990). The flavin-containing monooxygenase expressed in rabbit liver and lung are products of related but distinctly different genes. *J. Biol. Chem.*, *265*, 5855–5861.

Lee, H.-K., Moon, J.-K., Chang, C.-H., Choi, H., Park, H.-W., Park, B.-S., et al. (2006). Stereoselective metabolism of endosulfan by human liver microsomes and human cytochrome P450 isoforms. *Drug Metab. Dispos.*, *34*, 1090–1095.

Lee, H.-K., Moon, J.-K., Chang, C.-H., Choi, H., Park, H.-W., Park, B.-S., et al. (2007). Correction to "Stereoselective metabolism of endosulfan by human liver microsomes and human cytochrome P450 isoforms". *Drug Metab. Dispos.*, *35*, 829–830.

Lemoine, A., Williams, D. E., Cresteil, T., & Leroux, J. P. (1991). Hormonal regulation of microsomal flavin-containing monooxygenases: Tissue dependent expression and substrate specificity. *Mol. Pharmacol.*, *40*, 211–217.

Leoni, C, Buratti, F. M., & Testai, E. (2008). The participation of human hepatic P450 isoforms, flavin-containing monooxygenases and aldehyde oxidase in the biotransformation of the insecticide fenthion. *Toxicol. Appl. Pharmacol.*, *233*, 343–352.

Levi, P. E., & Hodgson, E. (1985). Oxidation of pesticides by purified cytochrome P-450 isozymes from mouse liver. *Toxicol. Lett.*, *24*, 221–228.

Levi, P. E., & Hodgson, E. (1988). Stereospecificity of the oxidation of phorate and phorate sulphoxide by purified FAD-containing monooxygenase and cytochrome P450. *Xenobiotica*, *1*, 29–39.

Levi, P. E., & Hodgson, E. (2001). Reactive metabolites and toxicity. In E. Hodgson & R. C. Smart (Eds.), *Introduction to biochemical toxicology* (pp. 199–220) (3rd ed.). New York: John Wiley & Sons.

Levi, P. E., Hollingworth, R. M., & Hodgson, E. (1988). Differences in oxidative dearylation and desulfuration of fenitrothion by cytochrome P-450 isozymes and in the subsequent inhibition of monooxygenase activity. *Pestic. Biochem. Physiol.*, *32*, 224–231.

Lewandowski, M., Chui, Y. C., Levi, P. E., & Hodgson, E. (1990). Differences in induction of hepatic cytochrome P450 isozymes in mice by eight methylenedioxyphenyl compounds. *J. Biochem. Toxicol.*, *5*, 47–55.

Li, A. A., Asbury, K. J., Hopkins, W. E., Feng, P. C. C., & Wilson, A. G. E. (1992). Metabolism of alachlor by rat and monkey liver and nasal turbinate tissues. *Drug Metab. Dispos.*, *20*, 616–618.

Li, H. C., Dehal, S. S., & Kupfer, D. (1995). Induction of the hepatic CYP2B and CYP3A enzymes by the proestrogenic pesticide methoxychlor and by DDT in the rat: Effects on methoxychlor metabolism. *J. Biochem. Toxicol.*, *10*, 51–61.

Lubet, R. A., Mayer, R. T., Cameron, J. W., Nims, R. W., Burke, M. D., Wolff, T., et al. (1985). Dealkylation of pentoxyresorufin: A rapid and sensitive assay for measuring induction of cytochromes P450 by phenobarbital and other xenobiotics in the rat. *Arch. Biochem. Biophys.*, *238*, 43–48.

Lubet, R. A., Guengerich, F. P., & Nims, R. W. (1990). The induction of alkoxyresorufin metabolism: A potential indicator of environmental contamination. *Arch. Environ. Contam. Toxicol.*, *19*, 157–163.

Lucier, G. W., & Menzer, R. E. (1970). Nature of oxidative metabolites of dimethoate formed in rats, liver microsomes and bean plants. *J. Agric. Food Chem.*, *18*, 698–704.

Lund, M. (1964). Resistance to warfarin in the common rat. *Nature*, *203*, 778.

Lund, M. (1967). Resistance of rodents to rodenticides. *World Rev. Pest Control*, *6*, 131–138.

Magdalou, J., & Hammock, B. D. (1987). Metabolism of tridiphane 2-3,5-dichlorophenyl-22,2,2-trichloroethyl oxirane by hepatic epoxide hydrolases and glutathione S-transferases in mouse. *Toxicol. Appl. Pharmacol.*, *91*, 439–449.

Mair, P., & Casida, J. E. (1991). Diallate, triallate and sulfallate herbicides: Identification of thiocarbamates, sulfoxidates, chloroacroleins and chloroallylthiols as mouse microsomal oxidase and glutathione S-transferase metabolites. *J. Agric. Food Chem.*, *39*, 1504–1508.

Markussen, M. D. K., Heiberg, A.-C., Fredholm, M., & Kristensen, M. (2008). Differential expression of cytochrome P450 genes between bromadiolone-resistant and anticoagulant susceptible Norway rats: A possible role for pharmacokinetics in bromadiolone resistance. *Pest Manage. Sci.*, *64*, 239–248.

Markussen, M. D. K., Heiberg, A.-C., Fredholm, M., & Kristensen, M. (2008). Identification of cytochrome P450 differentiated expression related to developmental stages in bromadiolone resistance in rats Rattus norvegicus. *Pestic. Biochem. Physiol.*, *91*, 147–152.

Matsumura, F. (1975). *Toxicology of insecticides*. New York: Plenum.

Matsumura, F. (1985). *Toxicology of insecticides* (2nd ed.). New York: Plenum.

Matthews, H. B., & Matsumura, F. (1969). Metabolic fate of dieldrin in the rat. *J. Agric. Food Chem.*, *17*, 845–852.

Mazur, C. S., Kenneke, J. F., Tebes-Stevens, C., Okino, M. S., & Lipscomb, J. C. (2007). In vitro metabolism of the fungicide and environmental contaminant trans-bromuconazole for risk assessment. *J. Toxicol. Environ. Health A Curr. Issues*, *70*, 1241–1250.

Mehendale, H. M., & Dorough, H. W. (1972). In A S Tahori (Ed.), *Insecticide–pesticide chemistry* (pp. 37–49). London: Gordon & Breach.

Mehmood, M., Kelly, D. E., & Kelly, S. L. (1996). Involvement of human cytochrome P450 3A4 in the metabolism of vamidothion. *Pestic. Sci.*, *46*, 287–290.

Mehmood, Z., Williamson, M. P., Kelly, D. E., & Kelly, S. L., (1996). Human cytochrome P450 3A4 is involved in the biotransformation of the herbicide 2,4-dichlorophenoxyacetic acid. *Environ. Toxicol. Pharmacol.*, *2*, 397–401.

Miksys, S., & Tyndale, R. F. (2009). Brain drug-metabolizing cytochrome P450 enzymes are active in vivo, demonstrated by mechanism-based enzyme inhibition. *Neuropsychopharmacology, 34*, 634–640.

Mitoma, C., & Udenfriend, S. (1962). Aryl-4-hydroxylase. *Methods Enzymol., 5*, 816–819.

Mortensen, A. S., & Arukwe, A. (2006). The persistent DDT metabolite, 1,1-dichloro-2,2-bisp-chlorophenylethylene, alters thyroid hormone-dependent genes, hepatic cytochrome P4503A, and pregnane X receptor gene expressions in Atlantic salmon *Salmo salar* parr. *Environ. Toxicol. Chem., 25*, 1607–1615.

Motoyama, N., & Dauterman, W. C. (1980). Glutathione S-transferases: Their role in the metabolism of organophosphorus insecticides. *Rev. Biochem. Toxicol., 2*, 49–69.

Mutch, E., Blair, P. G., & Williams, F. M. (1999). The role of metabolism in determining susceptibility to parathion toxicity in man. *Toxicol. Lett., 107*, 177–187.

Mutch, E., Daly, A. K., Leathart, J. B. S., Blain, P. G., & Williams, F. M. (2003). Do multiple P450 isoforms contribute to parathion metabolism in man? *Arch. Toxicol., 77*, 313–320.

Mutch, E., Daly, A. K., & Williams, F. M., (2007). The relationship between PON1 phenotype and PON1-192 genotype in detoxication of three oxons by human liver. *Drug Metab. Dispos., 35*, 315–320.

Neal, R. A., (1972). A comparison of the *in vitro* metabolism of parathion in the lung and liver of the rabbit. *Toxicol. Appl. Pharmacol., 23*, 123–130.

Neal, R. A. (1980). Microsomal metabolism of thiono-sulfur compounds: Mechanisms and toxicological significance. *Rev. Biochem. Toxicol., 2*, 131–171.

Neal, R. A., & Halpert, J. (1982). Toxicology of thiono-sulfur compounds. *Annu. Rev. Pharmacol. Toxicol., 22*, 321–339.

Neal, R. A., Sawahata, T., Halpert, J., & Kamataki, T. (1983). Chemically reactive metabolites as suicide enzyme inhibitors. *Drug Metab. Rev., 14*, 49–59.

Nelson, D. R. (2006). Cytochrome P450 nomenclature, 2004. *Methods Mol. Biol., 320*, 11.

Nomeir, A. A., & Dauterman, W. C. (1979). *In vitro* metabolism of EPN and EPNO by mouse liver. *Pestic. Biochem. Physiol., 10*, 190–196.

Oonithan, E. S., & Casida, J. E. (1966). Metabolites of methyl- and dimethyl-carbamate insecticide chemicals as formed by rat-liver microsomes. *Bull. Environ. Contam. Toxicol., 1*, 59–69.

Oonithan, E. S., & Casida, J. E. (1968). Oxidation of methyl and dimethylcarbamate insecticide chemicals by microsomal enzymes and anti-cholinesterase activity of the metabolites. *J. Agric. Food Chem., 16*, 28–44.

Ophof, A. J., & Langeveld, D. W. (1969). Warfarin-resistance in the Netherlands. *Schriftenr. Ver. Wasser-Boden-Lufthyg., 32*, 39.

O'Reilly, R. A., Aggeler, P. M., & Leong, L. S. (1963). Studies on the coumarin anticoagulant drugs: The pharmacodynamics of warfarin in man. *J. Clin. Invest., 4*, 1542–1551.

Ozburn, G. W., & Morrison, F. O. (1962). Development of a DDT-tolerant strain of laboratory mice. *Nature, 196*, 1006–1010.

Ozburn, G. W., & Morrison, F. O. (1965). The effect of DDT on respiratory metabolism of DDT-tolerant mice *Mus musculus. Can. J. Zool., 43*, 709–717.

Papadopoulos, N. M. (1964). Formation of nornicotine and other metabolites from nicotine *in vitro* and *in vivo. Can. J. Biochem., 42*, 435–442.

Parke, D. V. (1987). Activation mechanisms in chemical toxicity. *Arch. Toxicol., 60*, 5–15.

Pehkonen, S. O., & Zhang, Q. (2002). The degradation of organophosphorus pesticides in natural waters: A critical review. *Crit. Rev. Environ. Sci. Tech., 31*, 17–72.

Pekas, J. C., & Paulson, G. D. (1970). Intestinal hydrolysis and conjugation of a pesticidal carbamate *in vitro. Science, 170*, 77–78.

Pelz, H. J. (2007). Spread of resistance to anticoagulant rodenticides in Germany. *Int. J. Pest Manage., 53*, 281–284.

Perkins, E. J., El-Alfy, A., & Schlenk, D. (1999). *In vitro* sulfoxidation of aldicarb by hepatic microsomes of the channel catfish *Ictalurus punctatus. Toxicol. Sci., 48*, 67–73.

Peters, R. A. (1952). Lethal synthesis. *Proc. R. Soc. London Ser. B, 139*, 143–170.

Peters, R. A. (1963). *Biochemical lesions and lethal synthesis*. New York: Macmillan Co.

Pinkhas, J., Djaldetti, M., Joshua, H., Resnick, C., & DeVries, A. (1963). Sulfhemoglobinemia and acute hemolytic anemia with Heinz bodies following contact with a fungicide–zinc ethylene bisdithiocarbamate in a subject with glucose-6-phosphate dehydrogenase deficiency and hypocatalasemia. *Blood*, *21*, 484–494.

Poet, T. S., Wu, H., Kousba, A. A., & Timchalk, C. (2003). In vitro rat hepatic and intestinal metabolism of the organophosphate pesticides chlorpyrifos and diazinon. *Toxicol. Sci.*, *72*, 193–200.

Pohl, R. J., & Fouts, J. R. (1980). A rapid method for assaying the metabolism of 7-ethoxyresorufin by microsomal subcellular fractions. *Anal. Biochem.*, *107*, 150–155.

Price, R. J., Giddings, A. M., Scott, M. P., Walters, D. G., Capen, C. C., Osimitz, T. G., et al. (2008). Effect of pyrethrins on cytochrome P450 isoforms in cultured rat and human hepatocytes. *Toxicology*, *243*, 84–85.

Quinn, G. P., Axelrod, J., & Brodie, B. B. (1958). Species, strain, and sex differences in metabolism of hexobarbitone, amidopyrine, aminopyrine, and aniline. *Biochem. Pharmacol.*, *1*, 152–159.

Rao, J. V. (2006). Biochemical alterations in euryhaline fish, Oreochromis mossambicus exposed to sub-lethal concentrations of an organophosphorus insecticide, monocrotophos. *Chemosphere*, *66*, 1814–1820.

Reed, C. J. (1993). Drug metabolism in the nasal cavity—relevance to toxicology. *Drug Metab. Rev.*, *25*, 173–205.

Reponen, P., Abass, K., Mattila, S., & Pelkonen, O. (2010). Overview of the metabolism and interactions of pesticide in hepatic in vitro systems. *Int. J. Environ. Anal. Chem.*, *90*, 429–437.

Richter, R. J., & Furlong, C. E. (1999). Determination of paraoxonase (PON1) status requires more than genotyping. *Pharmacogenetics*, *9*, 745–753.

Richter, R. J., Jarvik, J. P., & Furlong, C. E. (2009). Paraoxonase 1 (PON1) status and substrate hydrolysis. *Toxicol. Appl. Pharmacol.*, *235*, 1–9.

Ripp, S. L., Itagaki, K., Philpot, R. M., & Elfarra, A. A. (1999). Species and sex differences in expression of flavin-containing monooxygenase form 3 in liver and kidney microsomes. *Drug Metab. Dispos.*, *27*, 46–52.

Rodriguez, C. J., & Harkin, J. M. (1995). New intermediates of dealkylation of ^{14}C atrazine by mouse liver microsomes. *Pestic. Biochem. Physiol.*, *53*, 23–33.

Ronis, M. J. J., & Cunny, H. C. (1994). Physiological endogenous factors affecting the metabolism of xenobiotics. In E. Hodgson & P. E. Levi (Eds.), *Introduction to biochemical toxicology* (pp. 133–151) (2nd ed.). Norwalk, CT: Appleton & Lange.

Ronis, M. J. J., & Cunny, H. C. (2000). Physiological endogenous factors affecting the metabolism of xenobiotics. In E. Hodgson & R. C. Smart (Eds.), *Introduction to biochemical toxicology* (3rd ed.). New York: Wiley.

Ronis, M. J. J., & Walker, C. H. (1989). The microsomal monooxygenases of birds. *Rev. Biochem. Toxicol.*, *10*, 301–384.

Rose, R. L., Tang, J., Choi, J., Cao, Y., Usmani, A., Cherrington, J., et al. (2005). Pesticide metabolism in humans, including polymorphisms. *Scand. J. Work. Environ. Health*, *31*(Suppl. 1), 156–163.

Ross, M. K., & Crow, J. A. (2007). Human carboxylesterases and their role in xenobiotic and endobiotic metabolism. *J. Biochem. Mol. Toxicol.*, *21*, 187–196.

Ross, M. K., Borazjani, A., Edwards, C. C., & Potter, P. M. (2006). Hydrolytic metabolism of pyrethroids by human and other mammalian carboxylases. *Biochem. Pharmacol.*, *71*, 657–669.

Sams, C., Mason, H. J., & Rawbone, R. (2000). Evidence for the activation of organophosphate insecticides by cytochrome P450 3A4 and 2D6 in human liver microsomes. *Toxicol. Lett.*, *116*, 217–221.

Sams, C., Cocker, J., & Lennard, M. S., (2004). Biotransformation of chlorpyrifos and diazinon by human liver microsomes and recombinant human cytochrome P450s (CYP). *Xenobiotica*, *34*, 861–873.

Sanchez, W., Piccini, B., & Porcher, J.-M. (2008). Effect of prochloraz fungicide on biotransformation enzymes and oxidative stress parameters in three-spined stickleback Gasterosteus aculeatus L. *J. Environ. Sci. Health B Pestic. Food Contam. Agric. Wastes*, *43*, 65–70.

Schenkman, J. B., Thummel, K. E., & Favreau, L. V. (1989). Physiological and patho-physiological alterations in rat hepatic cytochrome P450s. *Drug Metab. Rev.*, *20*, 557–584.

Schlenk, D., Cashman, D. R., Yeung, C., Zhang, X., & Rettie, A. E. (2002). Role of human flavin-containing monooxygenases in the sulfoxidation of [^{14}C]aldicarb. *Pestic. Biochem. Physiol.*, *73*, 67–73.

Schultz-Jander, D. A., & Casida, J. E. (2002). Imidacloprid insecticide metabolism: Human cytochrome P450 isozymes differ in selectivity for imidazolidine oxidation versus nitroimine reduction. *Toxicol. Lett.*, *132*, 65–70.

Shi, X., Dick, R. A., Ford, K. A., & Casida, J. E (2009). Enzymes and inhibitors in neonicotinoid insecticide metabolism. *J. Agric. Food Chem.*, *57*, 4861–4866.

Smith, C. A. D., Smith, G., & Wolf, C. R. (1994). Genetic polymorphisms in xenobiotic metabolism. *Eur. J. Cancer*, *31A*, 1921–1935.

Smith, C. A. D., Smith, G., & Wolf, C. R. (1994). Genetic polymorphisms in xenobiotic metabolism. *Eur. J. Cancer*, *30A*, 1935–1941.

Smyser, B. P., & Hodgson, E. (1985). Metabolism of phosphorus-containing compounds by pig liver microsomal FAD-containing monooxygenase. *Biochem. Pharmacol.*, *34*, 1145–1150.

Smyser, B. P., Sabourin, P. J., & Hodgson, E. (1985). Oxidation of pesticides by purified microsomal FAD-containing monooxygenase from mouse and pig liver. *Pestic. Biochem. Physiol.*, *24*, 368–374.

Smyser, B. P., Levi, P. E., & Hodgson, E. (1986). Interactions of diethylphenylphosphine with purified reconstituted mouse liver cytochrome P-450 monooxygenase systems. *Biochem. Pharmacol.*, *35*, 1719–1723.

Sonderfan, A. J., Arlotto, M. P., Dutton, D. R., McMillen, S. K., & Parkinson, A. (1987). Regulation of testosterone hydroxylation by rat liver microsomal cytochrome P450. *Arch. Biochem. Biophys.*, *255*, 27–41.

Speerschneider, P., & Dekant, W. (1995). Renal tumorigenicity of 1,1-dichloroethene in mice: The role of male-specific expression of cytochrome P450 2E1 in the renal bioactivation of 1,1-dihloroethene. *Toxicol. Appl. Pharmacol.*, *130*, 48–56.

Stangroom, S. J., Collins, C. D., & Lester, J. N. (2000). Abiotic behaviour of organic micropollutants in soils and the aquatic environment: A review, II. *Biotransform. Environ. Tech.*, *21*, 865–882.

Stok, J. E., Huang, H., Jones, P. D., Wheelock, C. E., Morisseau, C., & Hammock, B. D. (2004). Identification, expression, and purification of a pyrethroid-hydrolyzing carboxylesterase from mouse liver microsomes. *J. Biol. Chem.*, *279*, 29863–29869.

Street, J. C., & Blau, S. E. (1972). Oxychlordane: Accumulation in rat adipose tissue on feeding chlordane isomers or technical chlordane. *J. Agric. Food Chem.*, *20*, 395–397.

Strother, A. (1972). *In vitro* metabolism of methylcarbamate insecticides by human and rat liver fractions. *Toxicol. Appl. Pharmacol.*, *21*, 112–129.

Suzuki, T., Nakagawa, Y., Tayama, K., Taguchi, K., & Suga, T. (2001). Toxicity and effects of 2,6-di-tert-butyl-4-methylphenyl N-methylcarbamate terbutol on hepatic cytochrome P450 in F344 rats. *Arch. Toxicol.*, *75*, 555–561.

Tang, J., Cao, Y., Rose, R. L., Hodgson, E., Brimfield, A. A., Dai, D., et al. (2001). Metabolism of chlorpyrifos by human cytochrome P450 isoforms and human, rat and mouse liver microsomes. *Drug Metab. Dispos.*, *29*, 1201–1204.

Tang, J., Cao, Y., Rose, R. L., & Hodgson, E. (2002). In vitro metabolism of carbaryl by human cytochrome P450 and its inhibition by chlorpyrifos. *Chem. Biol. Interact.*, *141*, 229–241.

Tang, J., Usmani, K. A., Hodgson, E., & Rose, R. L. (2004). In vitro metabolism of fipronil by human and rat cytochrome P450 and its interactions with testosterone and diazepam. *Chem. Biol. Interact.*, *147*, 319–329.

Telle, H. J. (1971). *Resistance to warfarin of the brown rat R. Norvegicus in Germany*. Geneva: World Health Organization. Publication No. WHO/VBC/71.331.

Tseng, Y. L., & Menzer, R. E. (1974). Effect of hepatic enzyme inducers on the *in vivo* and *in vitro* metabolism of dicrotophos, dimethoate and phosphamidon in mice. *Pestic. Biochem. Physiol.*, *4*, 425–437.

Tyndale, R. F., & Sellers, E. M. (2002). Genetic variation in CYP2A6-mediated nicotine metabolism alters smoking behavior. *Ther. Drug Monit.*, *24*, 163–171.

Tynes, R. E., & Hodgson, E. (1983). Oxidation of thiobenzamide by the FAD-containing and cytochrome P450-dependent monooxygenases of liver and lung microsomes. *Biochem. Pharmacol.*, *32*, 3419–3428.

Tynes, R. E., & Hodgson, E. (1985a). Catalytic activity and substrate specificity of the flavin-containing monooxygenases in microsomal systems: Characterization of the hepatic, pulmonary and renal enzymes of the mouse, rabbit, and rat. *Arch. Biochem. Biophys.*, *240*, 77–93.

Tynes, R. E., & Hodgson, E. (1985b). Magnitude of involvement of the mammalian flavin-containing monooxygenase in the microsomal oxidation of pesticides. *J. Agric. Food Chem.*, *33*, 471–479.

Tynes, R. E., Sabourin, P. J., & Hodgson, E. (1985). Identification of distinct hepatic and pulmonary forms of microsomal flavin-containing monooxygenase in the mouse and rabbit. *Biochem. Biophys. Res. Commun.*, *126*, 1069–1075.

U.S. Environmental Protection Agency (1986). Guidance for carcinogenic risk assessment. *Fed. Register*, *51185*, 33992–34003.

U.S. Environmental Protection Agency (1987). *Alachlor. Notice of intent to cancel registrations: Conclusions of special review.* Washington, DC: Office of Pesticide Programs, Environmental Protection Agency.

Usmani, K. A., Rose, R. L., Goldstein, J. A., Taylor, W. G., Brimfield, A. A., & Hodgson, E. (2002). *In vitro* human metabolism and interactions of repellent N,N-diethyl-m-toluamide. *Drug Metab. Dispos.*, *30*, 289–294.

Usmani, K. A., Hodgson, E., & Rose, R. L. (2004a). *In vitro* metabolism of carbofuran by human, mouse, and rat cytochrome P450 and interactions with chlorpyrifos, testosterone, and estradiol. *Chem. Biol. Interact.*, *150*, 221–232.

Usmani, K. A., Karoly, E. D., Hodgson, E., & Rose, R. L. (2004b). *In vitro* sulfoxidation of thioether compounds by cytochrome P450 and flavin-containing monooxygenase isoforms with particular reference to the CYP2C subfamily. *Drug Metab. Dispos.*, *32*, 333–339.

Van Bezooijen, C. F. A. (1984). Influence of age-related changes in rodent liver morphology and physiology on drug metabolism—a review. *Mech. Ageing Dev.*, *25*, 1–22.

Van Bezooijen, C. F. A., Horbach, G. J. J. J., & Hollander, C. F. (1986). The effect of age on rat liver drug metabolism. In D. Platt (Ed.), *Drugs and aging* (pp. 45–55). Berlin: Springer-Verlag.

Venkatesh, K., Levi, P. E., & Hodgson, E. (1991). The flavin-containing monooxygenase of mouse kidney: A comparison with the liver enzyme. *Biochem. Pharmacol.*, *42*, 1411–1420.

Venkatesh, K., Levi, P. E., Inman, A. O., Monteiro-Riviere, N. A., Misra, R., & Hodgson, E. (1992). Enzymatic and immunohistochemical studies on the role of cytochrome P450 and the flavin-containing monooxygenase of mouse skin in the metabolism of pesticides and other xenobiotics. *Pestic. Biochem. Physiol.*, *43*, 53–66.

Venkatesh, K., Blake, B., Levi, P. E., & Hodgson, E. (1992). The flavin-containing monooxygenase in mouse lung: Evidence for expression of multiple forms. *J. Biochem. Toxicol.*, 7, 163–169.

Vinson, S. B., Boyde, C. E., & Ferguson, D. E. (1963). DDT resistance in the mosquito fish *Gambusia affinis*. *Science*, *139*, 217–218.

Walker, C. H. (1983). Pesticides and birds: Mechanisms of selective toxicity. *Agric. Ecosyst. Environ.*, *9*, 211–226.

Walker, C. H. (1994). Comparative toxicology. In E. Hodgson & P. E. Levi (Eds.), *Introduction to biochemical toxicology* (pp. 193–218) (2nd ed.). Norwalk, CT: Appleton & Lange.

Wang, H., Feng, Z., Yao, D., Sui, J., Zhong, W., Li, M., & Dai, J. Y. (2008). Warfarin resistance in Rattus losea in Guangdong province, China. *Pestic. Biochem. Physiol.*, *91*, 90–95.

Wattenberg, L. W. (1971). Studies of polycyclic hydrocarbon hydroxylases of the intestine possibly related to cancer: Effect of diet on benzpyrene hydroxylase activity. *Cancer Philadelphia*, *28*, 99–102.

Webb, R. E., & Horsfall, F., Jr. (1967). Endrin resistance in the pine mouse. *Science*, *156*, 1762.

Werle, E., & Uschold, E. (1948). Enzymatic detoxication of nicotine by animal tissue. *Biochem. Z.*, *318*, 531–537. [in German]

Werringloer, J. (1978). Assay of formaldehyde generated during microsomal oxidation reaction. In S. Fleischer & L. Packer (Eds.), *Methods in enzymology* (Vol. 52, pp. 297–302). New York: Academic Press.

Wheelock, C. E., Shan, G., & Ottea, J. (2005). Overview of carboxylases and their role in the metabolism of pesticides. *J. Pestic. Sci.*, *30*, 75–83.

Williams, D. E., Ziegler, D. M., Norden, D. J., Hale, S. E., & Masters, B. S. S. (1984). Rabbit lung flavin-containing monooxygenase is immunochemically and catalytically distinct from the liver enzyme. *Biochem. Biophys. Res. Commun.*, *125*, 116–122.

Williams, D. E., Hale, S. E., Muerhoff, A. S., & Masters, B. S. S. (1985). Rabbit lung flavin-containing monooxygenase: Purification, characterization, and induction during pregnancy. *Mol. Pharmacol.*, *28*, 381–390.

Wirth, P. J., & Thorgeirsson, S. S. (1978). Amine oxidase in mice—sex differences and developmental aspects. *Biochem. Pharmacol.*, *27*, 601–603.

Wood, A. W., Ryan, D. E., Thomas, P. E., & Levin, W. (1983). Regio- and stereoselective metabolism of two C19 steroids by five highly purified and reconstituted rat hepatic cytochrome P450 isozymes. *J. Biol. Chem.*, *258*, 8839–8847.

Wormhoudt, L. W., Cammandeur, J. N. M., & Vermeulen, N. P. E. (1999). Genetic polymorphisms of human N-acetyltransferase, cytochrome P450, glutathione-S-transferase, and epoxide hydrolase enzymes: Relevance to xenobiotic metabolism and toxicity. *Crit. Rev. Toxicol.*, *29*(59–124).

Wright, A. S., Potter, D., Wooder, M. F., Donninger, C., & Greenland, R. D. (1972). Effects of dieldrin on mammalian hepatocytes. *Food Cosmet. Toxicol.*, *10*, 311–322.

Xenotech. (2008). Donor information for human liver microsomes. Xenotech, Lenexa, KS, USA.

Xia, S. Y., Zhao, S. S., Wang, X. D., et al. (1995). Studies on the *in vitro* metabolism of 7-N,N-dimethylamino-1,2,3,4,5-pentathiocyclooctane by rat liver microsomes treated with phenobarbital. *Pestic. Biochem. Physiol.*, *51*, 48.

Yang, R. S. H., Dauterman, W. C., & Hodgson, E. (1969). Enzymatic degradation of diazinon by rat liver microsomes. *Life Sci.*, *8*, 667–672.

Yang, R. S. H., Hodgson, E., & Dauterman, W. C. (1971). Metabolism *in vitro* of diazinon and diazoxon in rat liver. *J. Agric. Food Chem.*, *19*, 10–13.

Yi, X. H., Ding, H., Lu, Y. T., Liu, H. H., Zhang, M., & Jiang, W. (2007). Effects of long-term alachlor exposure on hepatic antioxidant defense and detoxifying enzyme activities in crucian carp Carassius auratus. *Chemosphere*, *68*, 1576–1581.

Zampaglione, N., Jollow, D. G., Mitchell, J. R., Stripp, B., Hamrick, M., & Gillette, J. R. (1973). Role of detoxifying enzymes in bromobenzene-induced liver necrosis. *J. Pharmacol. Exp. Ther.*, *187*, 218–227.

Ziegler, D. M. (2002). An overview of the mechanism, substrate specificities, and structure of FMOs. *Drug Metab. Rev.*, *34*, 503–511.

Further Reading

Campbell, A, Holstege, D, Swezey, R, & Medina-Cleghorn, D. (2008). Detoxification of molinate sulfoxide: Comparison of spontaneous and enzymatic glutathione conjugation using human and rat liver cytosol. *J. Toxicol. Environ. Health A*, *71*, 1338–1347.

Coulet, M., Dacasto, M., Eeckhoutte, C., Larrieu, G., Sutra, J. F., & Alvinerie, M., et al. (1998). Identification of human and rabbit cytochromes P450 1A2 as major isoforms involved in thiabendazole 5-hydroxylation. *Fundam. Clin. Pharmacol.*, *12*, 225–235.

Davison, A. N. (1955). The conversion of Schradan OMPA and parathion into inhibitors of cholinesterase by mammalian liver. *Biochem. J.*, *61*, 203–209.

Mazur, C. S., & Kenneke, J. F. (2008). Cross-species comparison of conazole fungicide metabolites using rat and rainbow trout (Oncorhynchus mykiss) hepatic microsomes and purified human CYP 3A4. *Environ. Sci. Technol*, *42*, 947–954.

Mutch, E, & Williams, F. M. (2006). Diazinon, chlorpyrifos and parathion are metabolised by multiple cytochromes P450 in human liver. *Toxicology*, *224*, 22–32.

Scollon, E. J., Starr, J. M., Godin, S. J., DeVito, M. J., & Hughes, M. F. (2009). *In vitro* metabolism of pyrethroids pesticides by rat and human hepatic microsomes and cytochrome P450 isoforms. *Drug Metab. Dispos.*, *37*, 221–228.

Yang, D, Wang, X, Chen, Y, Deng, R, & Yan, B. (2009). Pyrethroid insecticides: Isoform-dependent hydrolysis, induction of cytochrome P450 3A4 and evidence on the involvement of the pregnane X receptor. *Toxicol. Appl. Pharmacol.*, *237*, 49–58.

CHAPTER 6

Distribution and Pharmacokinetics Models*

Ronald E. Baynes, Kelly J. Dix and Jim E. Riviere
North Carolina State University, Raleigh, NC, USA

OUTLINE

INTRODUCTION

For a pesticide to elicit toxicity, it must be transferred from the external site of exposure to the target site (e.g., organ, nucleic acid, receptor) and achieve a sufficiently high

---

*This chapter was editorially abbreviated and reformatted from K.J. Dix, 2001, Absorption, distribution and pharmacokinetics, Chapter 24 in *Handbook of Pesticide Toxicology* (R. Krieger, Ed.), 2nd edition, Academic Press, San Diego.

Pesticide Biotransformation and Disposition
DOI: 10.1016/B978-0-12-385481-0.00006-X

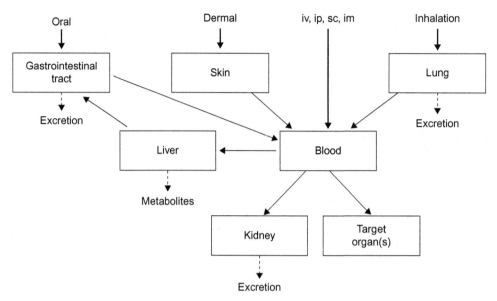

Figure 6.1 Representation of the absorption, distribution, metabolism, and excretion of toxicants.

concentration in the target organ (Figure 6.1). Absorption is the translocation of the pesticide from an external source of exposure to the bloodstream. Once in the blood, the chemical is distributed through the body and delivered to the tissues, where it may leave the blood and enter the cells of the tissue, or remain in the blood and simply pass through the tissue. In certain tissues, such as the liver, the chemical may be effectively removed from the body by metabolism. Other tissues, such as the kidney and lung, serve to eliminate xenobiotics from the body by excretion. Absorption, distribution, metabolism, and excretion, which are collectively termed disposition, are all factors that affect the concentration of a chemical in target tissues. Pharmacokinetics refers to the mathematical description of the time course of chemical disposition in the body. Metabolism and excretion are discussed in detail in other chapters of this work. This chapter focuses on chemical distribution and the basic pharmacokinetic modeling techniques used to mathematically describe chemical disposition.

DISTRIBUTION

Once in the bloodstream, the chemical is available for distribution throughout and elimination from the body. Metabolism and excretion, which are components of elimination, are discussed in other chapters. This section focuses on distribution, the *reversible* translocation of chemicals from one location to another in the body. Distribution of a toxicant to and accumulation in the target organ may result in toxicity. Accumulation at nontarget sites, on the other hand, results in storage of the pesticide away from the site of action and ultimately protection from toxicity. The physiology of the organism and the physicochemical characteristics of the pesticide are important factors in the distribution of absorbed pesticides.

Total Body Water

Chemicals in the body move throughout the water compartments of the body. As already discussed, the ability of chemicals to move between the various water compartments is limited by the physicochemical properties of the chemical. Total body water consists of plasma water, interstitial water, and intracellular water. In humans, approximately 60% of body weight is water, with plasma, intracellular, and interstitial water accounting for 5, 15, and 40% of body weight, respectively. Plasma water, which represents approximately 53–58% of blood volume in humans, plays an essential role in the distribution of absorbed chemicals.

For a chemical to move from blood (plasma water) into tissues, it must cross the endothelial cell layer lining the capillaries (i.e., capillary wall) to enter the interstitial water and then cross the plasma membrane to enter the intracellular water. Chemicals exist in blood as free circulating chemicals or are noncovalently bound to plasma proteins. The rates of association with and dissociation from plasma proteins are very rapid (on the order of milliseconds), and it is assumed that the bound and free forms of the chemical are in equilibrium. The capillary wall is permeable to small molecules, but not readily permeable to high-molecular-weight molecules such as plasma proteins. Only free chemicals that are small enough to pass through the capillaries, then, are available to move from plasma water to interstitial water. The processes for crossing plasma membranes described in Chapter 3 govern passage from the interstitial water to the intracellular water.

Rate and Extent of Distribution

Factors that influence the rate and extent of distribution of a chemical to a particular tissue include the blood flow to the tissue (rate of delivery), the mass of the tissue, the ability of the chemical to cross membranes, and the affinity of the chemical for the tissue relative to blood. The rate of distribution of a chemical from blood to tissues can be perfusion- or diffusion-rate limited. For lipophilic chemicals that rapidly cross membranes, the rate of delivery to tissues is limited by blood flow (perfusion-rate limited). For polar and ionized chemicals that do not readily cross the plasma membrane, the rate of delivery to tissues is limited by diffusion (diffusion-rate limited). Plasma protein binding increases the rate of distribution to tissues for toxicants that are not diffusion-rate limited. The free toxicant may readily cross the capillary wall, effectively decreasing its free concentration in blood. Bound toxicant then dissociates from plasma proteins to maintain the equilibrium between the bound and the free forms, yet the *new* free molecules rapidly leave the blood, which further increases the dissociation of bound toxicant, and so on. In contrast, distribution of more polar compounds that are diffusion-rate limited is dependent on the extent of protein binding.

Initial distribution is influenced primarily by blood flow to tissues, whereas final distribution is influenced primarily by the affinity of the chemical for various tissues relative to blood (i.e., the tissue partition coefficient). In the early phase of distribution, tissues that receive a high blood flow (e.g., liver, kidney, and brain) may achieve high concentrations of the chemical even though the tissue partition coefficient for that chemical is low. Likewise, tissues that are slowly perfused (e.g., adipose) may achieve a low concentration of the chemical in the early phase of distribution even though the tissue partition coefficient for that chemical is very high. Later in the distribution phase, however, the chemical redistributes to tissues based on tissue partition coefficients, and the chemical is more concentrated in tissues with relatively high partition coefficients. Pesticides and other xenobiotics do not have the same tissue partition coefficient for all tissues. For example, dimethoate has a relatively high affinity for liver, muscle, and brain (Garcia–Repetto et al., 1995), whereas the chlorinated insecticides DDT, aldrin, and dieldrin are lipophilic and have high affinities for adipose tissue (Lehman, 1956; Robinson et al., 1969).

Volume of Distribution

When a chemical is absorbed and distribution is complete, its concentration in blood depends on the amount absorbed and the extent of tissue distribution. The apparent volume of distribution (V_d) is a proportionality constant that relates the amount of chemical in the body to its concentration in plasma,

$$V_d = \frac{(\text{Amount in body})}{(\text{Concentration in plasma})} \quad (1)$$

where V_d is the *theoretical* volume of fluid the chemical would occupy to achieve the observed concentration in plasma and does not necessarily correspond to the volume of a particular body fluid compartment. For example, a chemical that is sequestered in a particular tissue will have a low concentration in plasma and a corresponding high volume of distribution, which may in fact be greater than the total body water. The V_d is often normalized by body weight, so the total body water is often quoted as 0.6 L/kg, and some chemicals whose distribution is greater than this physiological value are often very lipophilic chemicals or weak bases as they have a greater affinity for tissues than for plasma proteins. As only the plasma concentration and not the tissue concentrations for these chemicals can be measured in humans, a large volume of distribution is computed based on the known small plasma concentration and known amount in the body. Many of the pesticides are lipophilic substances and fall into this category of chemical with large volumes of distribution. However, chemicals that are large macromolecules and/or weak acids that bind to plasma proteins are less likely to display extravascular distribution, that is, move from the plasma into tissues. Such chemicals are described as having small volumes of distribution and their V_d values are less than the normal physiological volume described above (0.6 L/kg) and are closer to the plasma volume of 3 L (0.05 L/kg) or the albumen volume (7.5 L or 0.1 L/kg), suggesting that the chemical after absorption was restricted to the plasma compartment because of its large molecular size and/or by noncovalent binding to albumin. The herbicide 2,4-D is a good example of a weak organic chemical that binds to plasma proteins and has a low volume of distribution (0.27 L/kg) in humans (Timchalk, 2004).

There are different ways to calculate V_d and these derived values have different interpretations and applications. For example, when computing how much chemical is in the body from a plasma concentration one needs to consider whether the state of drug distribution is, at time 0, under equilibrium conditions, or at a pseudo-equilibrium state of distribution (Toutain and Bousquet-Melou, 2004a). The apparent volume of the central compartment, V_c, is for intravenous (iv) doses only and is calculated as follows:

$$V_c = \frac{D}{\sum C_n}, \tag{2}$$

where D is the dose and C_n are the intercepts of the various phases of chemical disposition obtained by fitting the concentration-time profile. V_c is the initial volume of distribution and it is related to the amount in the body (i.e., the dose, D) before the chemical is distributed or eliminated. If distribution is instantaneous, then the body can be reduced to one homogeneous compartment and V_c remains the V_d. However, there are many substances for which there is a non-instantaneous distribution, and there is a distribution phase during which the chemical partitions into various tissues, and the decline in plasma concentration is not due to elimination but to this partitioning process. This continues until pseudo-equilibrium of distribution is achieved, whereby the net exchange between the central compartment and the tissue compartment is zero. At this point further decline in plasma concentration is due to irreversible elimination of the chemical from the body. It is during this terminal phase that a V_d value known as V_{area}, which is greater than the initial V_d (i.e., V_c), can be calculated; it takes into account the amount of chemical in the body and the plasma concentration during this phase. The apparent volume of distribution, V_{area}, is based on trapezoid AUC_∞ (area) and elimination rate, λ_z. This can apply to iv but also to oral exposure if complete absorption ($F = 1$) is assumed and can be calculated as follows:

$$V_{area} = \frac{F D}{AUC_\infty}. \tag{3}$$

The third V_d value that may be reported is the apparent volume of distribution under steady-state conditions, V_{ss}. This can be estimated graphically from trapezoid total area measurements and applies to an iv infusion dose or to situations in which the clearance is apparently zero; that is, the rate of chemical input equals the rate of chemical output and thus V_{ss} is a volume term that is independent of clearance:

$$V_{ss} = \frac{D[AUMC_\infty]}{[AUC_\infty]^2}. \tag{4}$$

The determination of AUC and AUMC is discussed in more detail under Noncompartmental Models, statistical moment theory and noncompartmental analysis. With all things being equal regarding the total amount of chemical in the body, the plasma concentration at pseudo-equilibrium conditions will always be lower than that during steady state, therefore V_{area} tends to be higher than V_{ss}. The reader should be aware that V_c values have limited application – although they can help predict initial maximum plasma concentration after an iv bolus, and adverse health effects can be correlated with this concentration. V_{area} can be used to estimate the residual amount of a chemical during the elimination phase, and V_{ss} is more relevant as in clinical pharmacology for accurately predicting the loading needed to achieve an instantaneous desired plasma concentration before steady-state infusion.

Blood-Brain Barrier

The blood-brain barrier, which protects the central nervous system, is not an absolute barrier. 2,4-Dichlorophenoxyacetic acid (2,4-D), for example, has been measured in the rabbit brain after an iv dose (Kim et al., 1996). The tight junctions of the capillary endothelial cells and the surrounding glial cell processes are the main structural features that contribute to the low permeability of the blood-brain barrier. Chemicals that circulate in the blood must pass through the capillary endothelial cell membrane and the glial cell membrane to enter the interstitial fluid of the brain. The low protein content of the brain's interstitial fluid limits lipophilic chemicals that are tightly bound to plasma proteins. Chemical access to the brain, then, is limited to those species that are free (unbound), lipophilic, nonionized, and transported by specialized carrier systems, whereas ionized and highly plasma protein bound chemicals are excluded by the blood-brain barrier. Another barrier to brain access is the presence of an adenosine 5'-triphosphate-dependent multidrug-resistance protein, which transports intracellular chemicals back into the extracellular space. Since the blood-brain barrier is not fully developed at birth, the risk of toxicity from exposure to some chemicals is higher for newborns and young children than it is for adults.

Placental Transfer

Functions of the placenta include delivery of nutrients to the fetus, removal of fetal waste, and maternal/fetal blood gas exchange, which suggests that many chemicals move freely across the placental membrane. In the framework of distribution, the placenta is not a barrier to protect the fetus from exposure to toxicants. Rather, the placenta is a typical plasma membrane barrier that is permeable to lipophilic and non-ionized molecules that readily cross plasma membranes, thereby exposing the fetus to toxicants. 1,1-Dichloro-2,2-bis(*p*-chlorophenyl)ethylene (DDE) and 2,4-D have been demonstrated to cross the placenta in rats, rabbits, and bats (Kim et al., 1996; Sandberg et al., 1996; Thies and McBee, 1994), and prenatal human exposure to 2,4-D has been associated with mental retardation in offspring (Casey and Collie, 1984).

Storage and Redistribution

Chemicals may accumulate in body compartments because of protein binding, active transport processes, or high solubility in (i.e., affinity for) a particular tissue. These sites of accumulation can be considered storage depots. Because a chemical in any tissue compartment is in equilibrium with its free concentration in blood, storage is dynamic. Removal of free chemical from the body by metabolism or excretion shifts the equilibrium such that stored chemical is released.

Plasma Proteins

Plasma protein binding plays a very important role in chemical-induced toxicity. Displacement of one chemical by another from plasma proteins can have severe consequences. If the bound chemical is very toxic, its displacement results in a higher free concentration in plasma, which results in greater availability for distribution to its site of toxic action.

Fat

Adipose tissue is a storage depot for a number of highly lipophilic chemicals, including pesticides. Storage in adipose tissue may be considered a protective mechanism, in that the pesticide is stored in a nontarget tissue, thereby lowering its concentration at the site of toxic action. For example, the chlorinated insecticides DDT (Dale et al., 1962; Hayes et al., 1958), chlordane (Ambrose et al., 1953), hexachlorobenzene isomers (Davidow and Frawley, 1951), lindane (Ludwig et al., 1964), aldrin, and dieldrin (Robinson et al., 1969) are lipophilic and accumulate in fat. Upon dieting and starvation, fat is mobilized and the stored chemical is released, which results in a sudden increase in the blood concentration and availability of the pesticide for redistribution. As was described for plasma proteins, chemicals stored in adipose tissue may be displaced by other chemicals. Street (1964) demonstrated that DDT displaces dieldrin from its storage sites in rat adipose, yet methoxychlor does not affect dieldrin storage. Toxicity may be observed if the released chemical is redistributed to the target organ.

Other Tissues and Tissue Components

Sequestration in tissues (e.g., kidney and liver) may be due to interaction of chemicals with tissue macromolecules such as proteins and nucleic acids, which influences the affinity of a tissue for a given chemical (tissue partition coefficient). Bone tissue, for example, is a potential storage depot for heavy metals.

Storage with Repeated Exposures

Body burden is the term for the concentration (or amount) of chemical in the body at any given time, and the biological half-life of a chemical is the time required to reduce the concentration of the chemical in the body by one-half, in the absence of further intake. Many pesticides are water soluble and easily excreted, or are readily metabolized to more water-soluble compounds that are easily excreted. Lipophilic pesticides, such as the organochlorines, however, are stored in fat and are not easily removed from the body, and most people around the world carry a low body burden of organochlorine pesticides (Burgaz et al., 1994; Durham, 1969; Zatz, 1972). Repeated exposure to a chemical may result in cumulative storage and an increased body burden. If the interval between exposures is long relative to the biological half-life of the chemical, all or most of it will be removed from the body prior to subsequent exposure, and it is unlikely that the chemical will accumulate. If the interval between exposures is short relative to the biological half-life, however, there will be a residual body burden from the first exposure when the second exposure occurs, and so on, such that the chemical accumulates in the body.

Cumulative storage of a chemical upon repeated exposure continues until a steady state of storage is reached. Factors that influence storage include exposure level (dosage), time interval between exposures, duration of repeated exposures, interaction with other chemicals, age, sex, species, disease status, and nutritional status. A mathematical discussion of storage is presented later in this chapter.

PHARMACOKINETICS

Pharmacokinetics is the modeling and mathematical description of the time course of chemical disposition (absorption, distribution, metabolism, and excretion). Although urine and exhaled breath may be obtained from humans, blood is the only *tissue* that can be readily and repeatedly sampled in humans. Pharmacokinetic models typically describe the change in blood (or plasma) concentration of the chemical with time. There are two basic approaches to characterizing the pharmacokinetics of a chemical in the body: compartmental and noncompartmental. Compartmental pharmacokinetic models represent the body as discrete compartments with mathematical descriptions of the movement of a chemical between compartments, including the processes of absorption and elimination. Compartmental models may be subdivided into classical and physiologically based models. In contrast, the noncompartmental approach assumes no compartmentalization of the body and applies the trapezoidal rule for calculating the area under the plasma concentration–time curve to characterize a chemical's pharmacokinetics.

Noncompartmental Models

Noncompartmental models use statistical moment theory for the analysis of plasma concentration-time data. The area under the plasma concentration vs. time curve (AUC) is a measure of the total systemic exposure to the chemical. The AUC is the integral of the rate of change of concentration in plasma as a function of time:

$$AUC = \int_0^\infty C\,dt. \tag{5}$$

The first moment of the plasma concentration vs. time curve is the plasma concentration multiplied by time vs. time curve, and the area under the first moment curve (AUMC) is:

$$AUMC = \int_0^\infty tC\,dt. \tag{6}$$

In pharmacokinetic studies, plasma samples are not collected for an infinite time, but rather the collection period ends at some time T. AUC and AUMC may be approximated from time 0 to time T using the trapezoidal rule (Figure 6.2) and then can be extrapolated from time T to ∞ by:

$$\int_T^\infty C\,dt = \frac{C_T}{\beta}, \tag{7}$$

$$\int_T^\infty tC\,dt = \frac{TC_T}{\beta} + \frac{C_T}{\beta^2}. \tag{8}$$

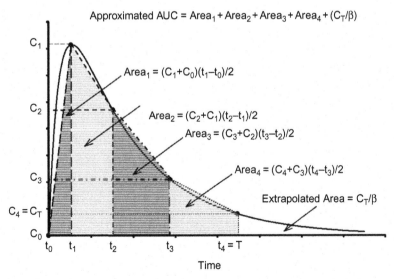

Figure 6.2 Determination of the area under the plasma concentration vs time curve using the trapezoidal rule.

In Eqs. (7) and (8), C_T is the observed concentration at the last time point T, and β is the slope of the terminal elimination phase of the log plasma concentration vs. time curve. The mean residence time (MRT) represents the overall persistence of the chemical in the body, and is thus the average time that the species stays in the body. It also represents the time required for 63.2% of the chemical to be eliminated from the body and can be calculated as:

$$\text{MRT} = \frac{\text{AUMC}}{\text{AUC}}. \tag{9}$$

The bioavailability (F), which is the fraction of chemical that is absorbed after extra-vascular administration, may also be calculated using noncompartmental models. Because the process of absorption is bypassed with intravenous administration, the bioavailability after intravenous administration is assumed to be unity. It is determined for extravascular administration (e.g., oral, dermal) with reference to an intravenous dose as:

$$F = \frac{D_{\text{iv}} \, \text{AUC}_{\text{ex}}}{D_{\text{ex}} \, \text{AUC}_{\text{iv}}}, \tag{10}$$

where D_{iv} and D_{ex} are the intravenous and extravascular doses, and AUC_{iv} and AUC_{ex} are the areas under the plasma concentration vs. time curve after intravenous and extravascular doses. When the intravenous and extravascular doses are the same, F is simply the proportion of AUC after extravascular and intravenous doses. Bioavailability is often referred to as a percentage. For example, the bioavailability of orally adminis-tered permethrin is 0.61, or 61% (Anadon et al., 1991). A number of factors, including the route of administration and species, may affect it. For example, the bioavailability of paraquat was 45, 12, or 3.8% after intratracheal, oral, or dermal doses, respectively (Chui et al., 1988). Species-dependent bioavailability has been shown for orally admin-istered metosulam, which was only 20% in mice and dogs, but greater than 70% in rats (Timchalk et al., 1996).

Plasma clearance (Cl), a measure of the inherent ability to remove a chemical from the body, is the volume of plasma that is cleared of the chemical per unit time. Cl after an intravenous dose is calculated as:

$$\text{Cl} = \frac{D_{\text{iv}}}{\text{AUC}}. \tag{11}$$

Cl can be calculated after an extravascular dose only if the bioavailability is 100% (i.e., $F = 1$).

The apparent volume of distribution at steady state (V_{ss} can be calculated after a single iv dose) is:

$$V_{\text{ss}} = \text{Cl} \times \text{MRT}_{\text{iv}}. \tag{12}$$

The first-order elimination rate constant (k_e) and elimination half-life ($t_{1/2}$) can also be calculated after a single iv dose for chemicals that appear to be characterized by a one-compartment model (see One-Compartment Model), from the relationships:

$$\text{MRT}_{iv} = \frac{1}{k_e} \tag{13}$$

and:

$$t_{1/2} = 0.693\,\text{MRT}_{iv}. \tag{14}$$

For chemicals that cannot be described by a one-compartment model,

$$\text{MRT}_{iv} = \frac{1}{k_e'}, \tag{15}$$

$$k_e' = \frac{Cl}{V_{ss}}, \tag{16}$$

and the *effective* elimination half-life is the product of 0.693 and MRT_{iv}.

Overview of Classical Compartmental Models

Classical compartmental models typically divide the body into one or more compartments that have no physiological or anatomical reality (Figure 6.3). It is assumed that the rate of transfer between compartments and the rate of elimination from the compartments are linear or first-order processes. Each model has an associated series of mathematical equations that describe the absorption, elimination, and transfer of chemicals between compartments. These equations are dependent only on the model structure and are independent of the chemical under study. Classical compartmental models can provide important parameters that describe chemical disposition, including the volume of distribution, absorption and elimination rate constants, elimination half-life, and plasma clearance. In the following discussion, elimination is assumed to occur only from compartment one, which is referred to as the central compartment.

One-Compartment Model

A one-compartment model (Figure 6.4), which represents the body as a single homogeneous compartment, adequately describes the pharmacokinetics of chemicals that rapidly equilibrate between blood and tissues. Therefore, it is reasonable to assume that the concentration of the chemical in blood (or plasma) is proportional to its concentration at the site of toxicity.

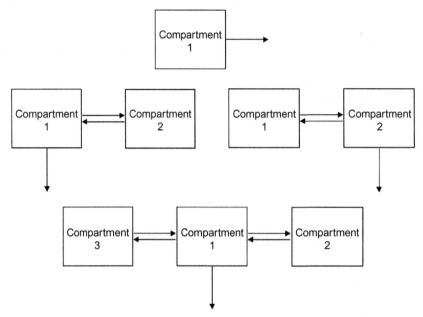

Figure 6.3 Schematic representations of one-, two-, and three-compartment models.

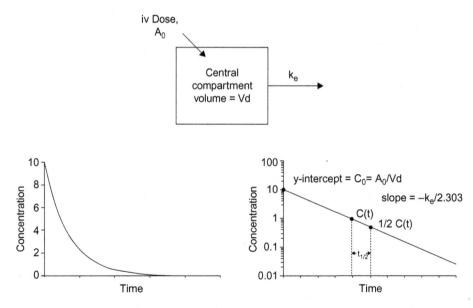

Figure 6.4 Representation of a one-compartment model with iv administration and first-order elimination, including a typical plasma concentration vs. time profile (linear and logarithmic scales). The volume of distribution, elimination rate constant, and elimination half-life are estimated by graphical methods.

Intravenous Bolus Dose

In the simplest one-compartment model, the chemical is introduced directly into the single compartment, and elimination occurs by a first-order process (Figure 6.4). The single compartment has a volume V_d, which in this case is the apparent volume of distribution. A typical plasma concentration-time curve for a one-compartment system is shown in Figure 6.4. This system is mathematically described by a first-order equation, in which the rate of removal of the chemical (mass per time) is proportional to the body load of the chemical (mass),

$$\frac{dA}{dt} = -k_e A, \tag{17}$$

where A is the amount of chemical in the body (units of mass) and k_e is the first-order elimination rate constant (units of reciprocal time), which represents the fractional elimination of chemical per unit time. A solution to Eq. (17) is:

$$A_t = A_0 \exp(-k_e t), \tag{18}$$

where A_t is the amount of chemical in the body at time t, and A_0 is the amount of chemical in the body at time 0. More frequently, the concentration rather than the amount of chemical is measured in plasma, and Eq. (18) is rewritten as:

$$C_t = C_0 \exp(-k_e t), \tag{19}$$

where C_t and C_0 are the concentrations (units of mass/volume) of the chemical in plasma at time t and time 0, respectively. Taking the logarithm of both sides of Eq. (19) yields:

$$\log C_t = \log C_0 \frac{k_e t}{2.303}. \tag{20}$$

The graph of $\log C_t$ vs. t has a y intercept of C_0 and a slope of $2k_e/2.303$; hence k_e can be determined from the slope of the $\log C_t$ vs. t graph (Figure 6.4). The apparent volume of distribution, V_d, can be determined from the known amount of chemical introduced into the body by intravenous injection at time 0 and the intercept of the $\log C_t$ vs. t graph as:

$$V_d = \frac{A_0}{C_0}. \tag{21}$$

The apparent volume of distribution is the volume into which the initial dose (A_0) would have to be dissolved to achieve the initial concentration of the chemical in plasma, C_0.

The elimination half-life ($t_{1/2}$) of a chemical is the time required for the amount or concentration in plasma to decrease by one-half, in the absence of additional exposure.

Therefore, C_t is equal to one-half of C_0 after one half-life has passed since the dose was administered, and:

$$\frac{C_0}{2} = C_0 \, exp(-k_e t_{1/2}).$$ (22)

Equation (22) can be solved for $t_{1/2}$,

$$t_{1/2} = \frac{0.693}{k_e},$$ (23)

which may also be estimated by inspection of the graph of log C_t vs. t (Figure 6.4). The reader should however be aware that the *terminal* half-life is not necessarily the time taken for the administered dose to decline by half, but rather the time required for plasma concentration to decline by half during the *terminal* phase of the concentration-time profile (Toutain and Bousquet-Melou, 2004b). The reader should compare Figure 6.4 (one-compartment model) and Figure 6.9 (two-compartment model) to see why this is important for the latter and why Eq. (23) is replaced with Eq. (52).

Plasma clearance (Cl, L/h) is a measure of the inherent ability to remove a chemical from the body and it is often normalized by body weight (L/h/kg). Cl represents the volume of plasma that is cleared of the chemical per unit time and is the ratio of the rate of elimination (mass/time) and concentration (mass/volume):

$$Cl = \left| \frac{dA/dt}{C_t} \right| = \frac{k_e A_t}{C_t} = \frac{k_e C_t V_d}{C_t} = k_e V_d.$$ (24)

Integration of Eq. (24) yields:

$$Cl = \frac{Dose}{AUC}.$$ (25)

Equation (24) can also be rearranged to solve for:

$$k_e = \frac{Cl}{V_d}.$$ (26)

Substitution of Eq. (25) into Eq. (26) and rearrangement leads to the equation for V_d:

$$V_d = \frac{Dose}{AUC \times k_e}.$$ (27)

Alternatively, the half-life can be calculated using V_{area} and Cl_{area} using iv data only as follows:

$$t_{1/2} = \frac{0.693 \times V_{area}}{Cl_{area}}.$$ (28)

Extravascular Dose

Humans are not typically exposed to pesticides by the intravenous route, but by extra-vascular routes (oral, dermal, inhalation), and the pesticide must be absorbed to enter the blood. Absorption is assumed to occur by a first-order process with an absorption rate constant k_a as shown in Figure 6.5. For extravascular exposure, then, the rate of removal of the chemical from the body is the net difference in the rates of introduction (by absorption) and elimination (by metabolism and excretion):

$$\frac{dA}{dt} = k_a A_a - k_e A. \tag{29}$$

In Eq. (29), A_a is the mass of chemical at the site of absorption and A is the mass of chemical in the body. As was noted earlier, extravascular exposure to chemicals is different from intravenous exposure, in that it cannot be assumed that 100% of the dose is absorbed. Some fraction F of the dose (D) is absorbed, or only the product FD is bio-available. The rate of removal of the chemical from the site of absorption is:

$$\frac{dA}{dt} = -k_a A_a. \tag{30}$$

Solving for A as a function of time in the preceding equations yields:

$$A_t = k_a FD \frac{\exp(-k_e t) - \exp(-k_a t)}{k_a - k_e}, \tag{31}$$

which can be rewritten in terms of concentration to yield:

$$C_t = \frac{k_a FD}{V_d} \frac{\exp(-k_e t) - \exp(-k_a t)}{k_a - k_e}. \tag{32}$$

A typical plasma concentration-time curve for a compound that is absorbed by a first-order process, rapidly equilibrates between blood and tissues and is eliminated by a first-order process is shown in Figure 6.5. After oral administration to rats, the plasma concentration vs. time profiles for triclopyr (Timchalk et al., 1996), diazinon (Wu et al., 1996), and paraquat (Chui et al., 1988) are all described by the model in Figure 6.5. Some time after administration, absorption is essentially complete and Eq. (32) is reduced to:

$$C_t = \frac{k_a FD}{V_d} \frac{\exp(-k_e t)}{k_a - k_e}. \tag{33}$$

Taking the logarithm of both sides of Eq. (33) yields:

$$\log C_t = \log \frac{k_a FD}{V_d (k_a - k_e)} = \frac{k_e t}{2.303}. \tag{34}$$

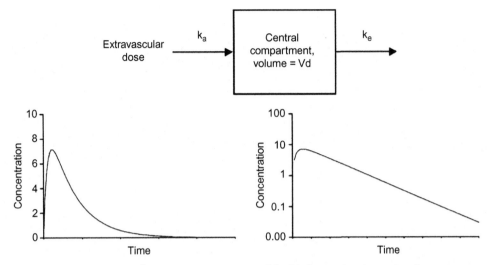

Figure 6.5 Representation of a one-compartment model with first-order absorption (i.e., extravascular administration) and first-order elimination, including a typical plasma concentration vs. time profile (linear and logarithmic scales).

The post–absorption phase of the graph of log C_t vs. t has a slope of $2k_e/2.303$, and as in the case of a one-compartment model with an intravenous dose, k_e can be determined from its terminal slope (Figure 6.6). The absorption rate constant may be obtained from the y intercept of the plasma concentration vs. time graph, which is $k_a FD/(V_d(k_a - k_e))$, or by the method of residuals as shown in the example in Figure 6.6 and Table 6.1. Integration of Eq. (34) from time 0 to infinity yields:

$$AUC = \frac{k_a FD}{V_d(k_a - k_e)} - \left(\frac{1}{k_e} - \frac{1}{k_a}\right), \tag{35}$$

which reduces to:

$$AUC = \frac{FD}{V_d k_e}. \tag{36}$$

Cl and V_d are derived from Eqs. (26) and (27), in which dose is adjusted for bioavailability:

$$Cl = \frac{FD}{AUC}, \tag{37}$$

$$V_d = \frac{FD}{AUC \times k_e}. \tag{38}$$

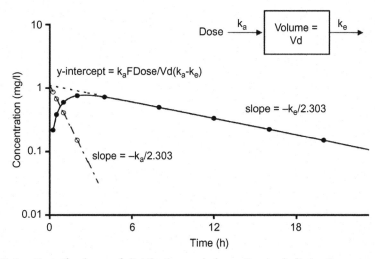

Figure 6.6 Estimation of volume of distribution and absorption and elimination rate constants for the one-compartment model in Figure 6.5 by graphical methods (i.e., curve stripping). Data are shown in Table 6.1.

Table 6.1 Data Used for the Method of Residuals Example Shown in Figure 6.6

Time (h)	Plasma	Extrapolated concentration (mg/L plasma)	Residual
0.25	0.218	1.083	0.865
0.5	0.382	1.057	0.675
1	0.597	1.005	0.408
2	0.759	0.910	0.151
4	0.724		
8	0.499		
12	0.334		
16	0.224		
20	0.150		
24	0.101		

Storage with Repeated Exposures

Repeated exposure to a chemical at constant time intervals may lead to accumulation of the chemical in the body until a steady state is achieved. During any exposure interval τ at steady state, the rate of chemical entry into the body is equal to the rate of its elimination (i.e., amount absorbed equals amount eliminated). Wagner (1967) proposed the concept of a concentration index (R_C), which provides information with regard to the increased accumulation over multiple exposures. R_C is defined as the ratio of the average concentration of a chemical in blood during an exposure interval of length τ at steady state $\bar{C}_\infty$ and the average concentration in blood during the same time interval after a single exposure $(\bar{C})$,

$$R_C = \frac{\bar{C}_\infty}{\bar{C}},$$ (39)

where $\bar{C}_\infty$ and $\bar{C}$ are defined as:

$$\bar{C}_\infty = \frac{1}{\tau} \int_{t_1}^{t_2} C_\infty dt,$$ (40)

$$\bar{C} = \frac{1}{\tau} \int_0^T C\, dt.$$ (41)

In Eq. (40), C is the concentration of the chemical in blood or plasma at time t after dosing during steady state, and $\tau = t_2 - t_1$. When C and C_∞ are measured, $\bar{C}_\infty$ and $\bar{C}$ can be calculated from the respective concentration vs. time curves using the trapezoidal rule.

Repeated administration of a given dose (D) of a chemical at fixed time intervals (τ) eventually leads to a steady state (equilibrium) of storage (Figure 6.7). The following discussion continues to assume that first-order processes govern absorption and elimination. Accumulation of a chemical in the body is described by the concentration index defined in Eq. (39). Wagner and colleagues (Wagner et al., 1965; Wagner, 1967)

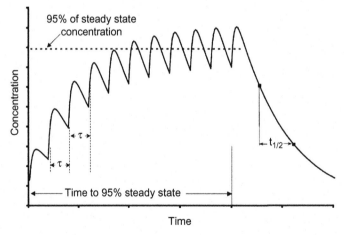

Figure 6.7 Approach to steady-state plasma concentration with repeated administration at constant dose intervals (τ). See Storage with Repeated Exposures for calculation of pharmacokinetic parameters.

derived the following equation for the average plasma concentration during any interval τ at steady state:

$$\bar{C}_\infty = \frac{FD}{V_d k_e \tau} = \frac{FD}{\text{Cl}\,\tau}. \tag{42}$$

The average plasma concentration during the first dose interval (from time 0 to τ) is:

$$\bar{C} = \frac{k_a FD}{V_d \tau (k_a - k_e)} \times \left(\frac{1 - \exp(-k_e \tau)}{k_e} - \frac{1 - \exp(-k_a \tau)}{k_a} \right). \tag{43}$$

Substituting Eqs. (42) and (43) into Eq. (39) and rearranging yields the concentration index:

$$R_C = \frac{1}{\left(1 - \left(\dfrac{k_a}{k_a - k_e} \exp(-k_e \tau) - \dfrac{k_e}{k_a - k_e} \exp(-k_a \tau) \right) \right)}. \tag{44}$$

For chemicals with $k_a \gg k_e$, which is the case for many of the organochlorine pesticides, for instance Zatz (1972),

$$\bar{C} = \frac{FD}{V_d k_e \tau}(1 - \exp(-k_e \tau)), \tag{45}$$

and:

$$R_C = \frac{1}{1 - \exp(k_e \tau)}. \tag{46}$$

The accumulation ratio (R_A) for multiple exposures at fixed time intervals was defined by Wagner (1967) as the ratio of the average mass of chemical in the body during any exposure interval at equilibrium, and the average mass of chemical absorbed after a single exposure. The average mass of chemical absorbed after a single exposure is simply FD, the percentage of dose absorbed. Use of the relationship between concentration, mass, and volume of distribution shown in Eq. (21) allows Eq. (42) to be rearranged to solve for the average mass of chemical during an exposure interval at steady state:

$$\bar{A}_\infty = \frac{FD}{k_e \tau}. \tag{47}$$

Substitution of Eq. (23) into Eq. (47) yields:

$$\bar{A}_\infty = \frac{1.44 FD t_{1/2}}{\tau}. \tag{48}$$

Hence,

$$R_A = \frac{1.44 FD t_{1/2}}{FD\tau} = \frac{1.44 t_{1/2}}{\tau}. \tag{49}$$

Multicompartment Models

Many chemicals do not rapidly equilibrate between blood and tissues, and their plasma concentration-time profiles do not conform to the one-compartment model already described. Instead, elimination from plasma is multiphasic, the simplest case being biphasic elimination. The early phase is referred to as the distribution phase, and the later phase is the postdistribution or elimination phase. The plasma concentration of the chemical declines more rapidly during the distribution phase than in the elimination phase. Two schematics of two-compartment models are shown in Figure 6.3. The central compartment includes blood and tissues in which the chemical rapidly equilibrates (e.g., tissues that receive a high blood flow) and is considered a homogeneous compartment. This is analogous to the single compartment of a one-compartment model. The peripheral compartment consists of tissues for which equilibrium is not instantaneous. In classical compartmental models, the chemical moves between the central and the peripheral compartments with associated transfer rate constants, but elimination is assumed to occur only from the central compartment with an associated elimination rate constant (Figure 6.8). Absorption, distribution, and elimination are assumed to be first-order processes.

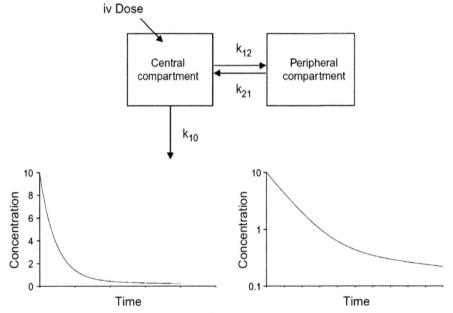

Figure 6.8 Representation of a two-compartment model with iv administration and first-order elimination, including a typical plasma concentration vs. time profile (linear and logarithmic scales).

Intravenous Bolus Dose

A schematic of a two-compartment model with first-order elimination is shown in Figure 6.8. The diazinon plasma concentration vs. time after an iv dose is represented by this two-compartment model (Wu et al., 1996). The concentration of chemical in plasma after an intravenous bolus dose as a function of time can be expressed as the sum of two monoexponential terms,

$$C_t = A_e^{-\alpha t} + B_e^{-\beta t}. \tag{50}$$

At some point after dosing, the distribution phase is complete and the only process that contributes to removal of the chemical from plasma is elimination. During this time, Eq. (50) reduces to:

$$C_t = B_e^{-\beta t}, \tag{51}$$

where B is the intercept and β is the slope of the *terminal* phase of the $\log|C_t$ vs. t curve (Figure 6.9). The rate constant β is analogous to k_e in the one-compartmental model described earlier. The elimination half-life is estimated from β according to the equation:

$$t_{1/2} = \frac{0.693}{\beta}. \tag{52}$$

The method of residuals is used to estimate A and α (Figure 6.9 and Table 6.2). The initial concentration, C_0, is determined by substituting $t = 0$ into Eq. (50):

$$C_0 = A + B. \tag{53}$$

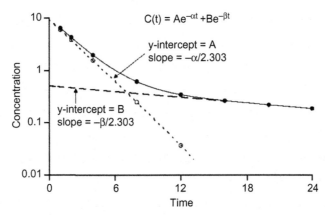

Figure 6.9 Estimation of volume of A, B, α, and β for the two-compartment model in Figure 6.10 by graphical methods (i.e., curve stripping). Data are shown in Table 6.2.

Table 6.2 Data Used for the Method of Residuals Example Shown in Figure 6.9

Time (h)	Plasma	Extrapolated concentration (mg/L plasma)	Residual
1	6.534	0.488	6.046
2	4.318	0.468	3.850
4	1.988	0.431	1.558
8	0.614	0.364	0.250
12	0.345	0.308	0.037
16	0.264		
20	0.220		
24	0.187		

Similar to the one-compartment model, the volume of the central compartment is:

$$V_c = \frac{\text{Dose}}{A + B}. \tag{54}$$

The rate constants α and β are composites of k_{12}, k_{21}, and k_{10} with the relationships:

$$\alpha + \beta = k_{12} + k_{21} + k_{10}, \tag{55}$$

$$\alpha\beta = k_{10}k_{21}. \tag{56}$$

The rate constants k_{12}, k_{21}, and k_{10} are determined from the relationships below (see Gibaldi and Perrier, 1982, for derivations):

$$k_{21} = \frac{A\alpha + B\beta}{A + B}, \tag{57}$$

$$k_{10} = \frac{-\beta}{k_{21}}, \tag{58}$$

$$k_{12} = \alpha = \beta - k_{21} - k_{10}. \tag{59}$$

Extravascular Dose

A schematic of a two-compartment model with first-order absorption and elimination is shown in Figure 6.10. Absorption and elimination are assumed to occur via the central compartment only. The plasma concentration as a function of time can be expressed as:

$$C_t = A\exp(-\alpha t) + B\exp(-\beta t) + C\exp(-k_{01}t). \tag{60}$$

It is often difficult to distinguish the absorption phase from the distribution phase of the log C_t vs. t curve, because k_{01} is similar in magnitude to α. Estimation of rate constants after an extravascular dose often requires data after an intravenous dose to distinguish between k_{01} and α (Gibaldi and Perrier, 1982).

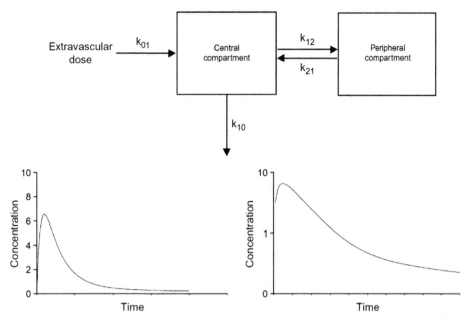

Figure 6.10 Representation of a two-compartment model with first-order absorption (i.e., extravascular administration) and first-order elimination, including a typical plasma concentration vs time profile (linear and logarithmic scales).

Physiologically Based Models

Unlike classical compartmental models, physiologically based pharmacokinetic (PBPK) models represent physiological and anatomical reality (Figure 6.11). The compartments are connected by blood flow, and chemicals may enter the body by any route. The model in Figure 6.11 incorporates exposure by the oral, dermal, and inhalation routes and elimination by urinary excretion, exhalation, and metabolism. PBPK models use mathematical descriptions of chemical disposition that are based on the physiological, physicochemical, and biochemical determinants of disposition, which include biochemical reaction rates and tissue partition coefficients for the chemistry and physiology (e.g., organ volumes, blood flows, respiration rates) of the animal. PBPK models do not assume that all processes governing disposition are linear, and saturable metabolism (Michaelis–Menten kinetics), for example, can be easily incorporated into them. PBPK models exist for pesticides from a variety of chemical classes (Table 6.3).

Like classical compartmental models, the compartments in Figure 6.11 represent organs or tissue groups in which a chemical is uniformly distributed, and arrows represent the pathways that govern chemical disposition. Each compartment has an associated volume. Absorption is represented by arrows to the portals of entry for various routes of exposure, blood flow is represented by arrows that interconnect the compartments of the model, and metabolism and excretion are represented by arrows from

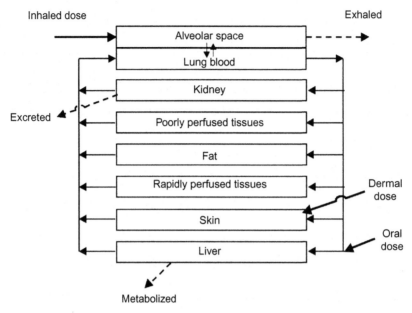

Figure 6.11 Schematic representation of a physiologically based pharmacokinetic model. This model contains descriptions of exposure by the inhalation, oral, and dermal routes and elimination by exhalation, excretion, and metabolism.

Table 6.3 Some Existing PBPK Models for Pesticides from Various Chemical Classes

Pesticide	References
Dieldrin	Leung and Paustenbach (1988)
Kepone (chlordecone)	el-Masri et al. (1995, 1996); Yang et al. (1995a,b)
Lindane (hexachlorocyclohexane)	DeJongh and Blaauboer (1997)
Hexachlorobenzene	Freeman et al. (1989); Roth et al. (1993)
Diisopropylfluorophosphate	Gearhart et al. (1994)
Dichlorobenzene	Hissink et al. (1997)
2,4-Dichlorophenoxyacetic acid	Kim et al. (1994, 1995, 1996)
Captan	Fisher et al. (1992); Woollen (1993)
Chlorpyrifos	Timchalk et al. (2002)

the compartments in which these processes occur. PBPK models require three types of parameters as inputs: physiological, physicochemical, and biochemical. Physiological parameters (e.g., pulmonary ventilation rate, cardiac output, blood flow to tissues, and tissue volumes) for humans and several laboratory animal species are available in the literature (Arms and Travis, 1988). Physiological parameters, which are not dependent on the chemical under study, are assumed to be constant for a given species. However, if it is known that physiological parameters change with time or a particular exposure scenario, those changes can be easily incorporated into PBPK models. Physicochemical parameters used in PBPK models (i.e., tissue partition coefficients) describe the relative solubility of the chemical in various media (e.g., air, blood, and tissues). Biochemical parameters include the rates of absorption, metabolism, macromolecular binding, and excretion. One of the major drawbacks of developing PBPK models is the large amount of physiological and biochemical data set requirements.

To develop a PBPK model, one must first consider which compartments of the organism to include. The compartments may be specific organs, anatomical regions, or lumped tissue groups, and their inclusion depends on which animal is being studied and whether the compartment contributes to the uptake, disposition, and/or toxicity of the chemical being modeled. For example, the lung, gastrointestinal tract, and skin may be included because of their ability to serve as sites of absorption, and the kidneys may be included because of their ability to serve as portals of excretion. Tissue partition coefficients and the metabolic capacity of a particular tissue also contribute to chemical disposition. For example, fat is often included in PBPK models because of the high partition coefficient of lipophilic chemicals (e.g., organochlorine pesticides) in adipose tissue, and the liver is often included as a separate compartment because of its involvement in the metabolism of a wide variety of chemicals. Tissues that are target sites for toxicity are also often included in PBPK models.

The next step in PBPK model development is to write a mass balance equation for each compartment to describe the rate of change of chemical concentration in that compartment as a function of time. In the most general case, the mass balance equation for each tissue compartment is:

$$\text{Rate of change} = (\text{Rate of uptake}) - (\text{Rate of removal}). \tag{61}$$

The rate of removal is the summation of removal by efflux back into the bloodstream, metabolism within the tissue, and excretion (e.g., biliary excretion in the liver or urinary excretion in the kidney). For simplicity, tissue compartments that have *no* capacity for metabolism or excretion will be considered. In this case, the rate of change of chemical concentration in the tissue is the difference in the rates of uptake and efflux. For uptake to occur from blood into a particular tissue compartment, the free chemical must diffuse out of the capillary space into the interstitial fluid and then diffuse across the plasma membrane to enter the intracellular space (Figure 6.12A). It is assumed that diffusion from the capillary membrane into the interstitial space is very rapid relative to diffusion from the interstitial space into the intracellular space, and the vascular and interstitial subcompartments are represented as one homogeneous subcompartment referred to as the extracellular space (Figure 6.12B). Free chemical in the blood enters the extracellular space of the tissue compartment at a rate (mass/time) that is the product of blood flow to the tissue (Q_t, units of volume/time) and the concentration of the free chemical in the arterial blood (C_a). Diffusion of the chemical from the extracellular space across the plasma membrane and into the intracellular space is governed by Fick's

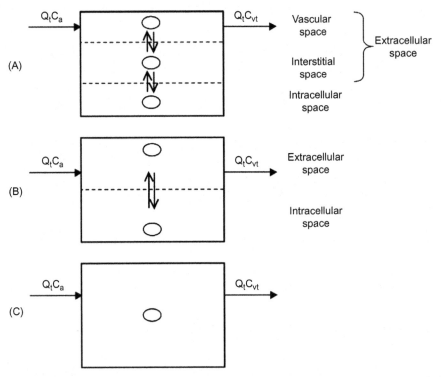

Figure 6.12 Uptake from the vascular space of a tissue compartment into the intracellular space where the tissue is represented as (A) three distinct compartments, (B) two compartments, or (C) a single homogeneous compartment.

law of diffusion, which states that the rate of transfer of chemical across a membrane (flux) is proportional to its concentration gradient across the membrane,

$$\text{Flux} = PA_t \Delta C, \tag{62}$$

where PA_t is the tissue membrane permeation area cross product and ΔC is the concentration gradient (units of mass/volume) of free chemical across the membrane.

For chemicals that have a perfusion-rate-limited distribution, diffusion of the chemical across the membrane is very rapid relative to its rate of delivery to the tissue (i.e., $PA_t \gg Q_t$), and the rate of uptake by tissues is limited by blood flow rather than the rate of diffusion across the membrane. For such chemicals, the free chemical concentration in the intracellular and extracellular spaces is in equilibrium, and the tissue compartment can be represented as a single homogeneous compartment as shown in Figures 6.11 and 6.12C. The mass balance equation for such a tissue compartment is:

$$V_t \frac{dC_t}{dt} = Q_t(C_a - C_{vt}), \tag{63}$$

where V_t is the volume of the tissue compartment, C_t is the concentration of free chemical in the tissue compartment, C_a is the concentration of free chemical in the entering arterial blood, and C_{vt} is the concentration of free chemical in the venous blood exiting the tissue. The concentrations of free chemical in tissue and venous blood leaving the tissue are related by the tissue-to-blood partition coefficient (P_t) as:

$$C_{vt} = \frac{C_t}{P_t}. \tag{64}$$

The overall mixed venous blood concentration is:

$$C_v = \frac{\sum C_{vt} Q_t}{Q_c}, \tag{65}$$

where Q_c is cardiac output (i.e., total blood flow or ϵQ_t).

For chemicals that have a distribution that is limited by the rate of diffusion across the cell membrane rather than by blood flow (i.e., diffusion-rate limited), separate mass balance equations must be written for the intracellular and extracellular compartments of the tissue (Figure 6.12B). The mass balance equation for the extracellular space is:

$$V_{es} \frac{dC_{es}}{dt} = Q_t (C_a - C_{vt}) + PA_t \left(\frac{C_t}{P_t} - C_{vt} \right), \tag{66}$$

where V_{es} and C_{es} are the volume of and the concentration in the extracellular space, respectively. The mass balance equation for the intracellular space (i.e., tissue matrix) is:

$$V_{is} \frac{dC_{is}}{dt} = PA_t \left(C_{vt} - \frac{C_t}{P_t} \right), \tag{67}$$

where V_{is} and C_{is} are the volume of and the concentration in the intracellular space, respectively.

Chemicals may be effectively eliminated from a tissue by metabolism, macromolecular binding, and/or excretion. For these tissues, the mass balance equation is more complex than those shown in Eqs. (63), (66), and (67). The mass balance equation for a tissue that metabolizes the chemical (e.g., the liver) is:

$$V_t \frac{dC_t}{dt} = Q_t (C_a - C_{vt}) - \frac{dA_{met}}{dt}, \tag{68}$$

where dA_{met}/dt is the rate of metabolism. Many enzyme systems are saturable, and:

$$\frac{dA_{met}}{dt} = \frac{V_{max} C_{vt}}{K_m + C_{vt}}, \tag{69}$$

where V_{max} and K_m are the maximal velocity and Michaelis constant of the enzymatic reaction. First-order metabolism is described by the equation:

$$\frac{dA_{met}}{dt} = V_t K_f C_{vt}, \tag{70}$$

where K_f is the first-order rate constant. Both saturable and first-order metabolism may occur simultaneously. For example, if metabolism by both saturable and first-order processes is occurring in the liver, the mass balance equation for the liver would be:

$$V_1 \frac{dC_1}{dt} = Q_1(C_a - C_{v1}) - \frac{V_{max}C_{v1}}{K_m + C_{v1}} - K_f C_{v1} V_1. \tag{71}$$

For inhalation exposure to volatile chemicals, equilibrium is established between the chemical in the alveolar air space of the lung and the chemical in arterial blood. The concentration of chemical in arterial blood (C_a) is described by the equation:

$$C_a = \frac{Q_c C_v + Q_p C_i}{Q_c + (Q_p/P_b)}, \tag{72}$$

where Q_p is alveolar ventilation rate, C_i is the concentration in inhaled air, and P_b is the blood:air partition coefficient. If the chemical can be inhaled, it can also be exhaled. The concentration of the chemical in exhaled alveolar air is the ratio of its concentration in arterial blood and the blood:air partition coefficient, C_a/P_b.

Upon oral ingestion of a chemical (e.g., food, drinking water), the chemical may be absorbed into the portal circulation or into the lymphatic system. For simplicity, we assume first-order absorption into the portal circulation only. In this situation, the chemical is delivered directly to the liver prior to distribution throughout the body. This requires an input term in the mass balance equation for the liver and a mass balance equation that describes the rate of loss of chemical from the site of absorption (stomach):

$$\frac{dA_{st}}{dt} = K_a A_{st}, \tag{73}$$

$$V_1 \frac{dC_1}{dt} = Q_1(C_a - C_{v1}) + K_a A_{st} - \frac{V_{max}C_{vt}}{K_m + C_{vt}} - K_f C_{v1} V_1, \tag{74}$$

where K_a is the oral absorption rate constant and A_{st} is the amount (mass) of chemical in the stomach (i.e., the site of absorption). Incorporation of dermal absorption is more complex, in that a skin compartment must be added to the model (see Krishnan and Andersen, 1994).

In addition to metabolism, chemicals may be eliminated from the body by excretion in urine, exhaled air, and other routes that are not discussed here (e.g., sweat, bile, and milk). A typical equation for the concentration of chemical in exhaled air is:

$$C_{\mathrm{ex}} = 0.7 \frac{C_{\mathrm{a}}}{P_{\mathrm{b}}} + 0.3C_1. \tag{75}$$

Equation (69) uses the assumption that exhaled air is a mixture of inhaled air (30%) and expired alveolar air (70%). Urinary excretion may be described in a number of ways, including excretion by first-order and saturable processes similar to the descriptions of metabolism in Eqs. (69) and (70). The same is true for elimination by other routes.

CONCLUSIONS

To either exert a deleterious effect or be detoxified, xenobiotics, including pesticides, must be absorbed, distributed, and/or metabolized; react with a macromolecule or other receptor; and/or be excreted. Distribution and pharmacokinetics are the subjects of this chapter. In the case of toxicants such as pesticides, the term toxicokinetics is frequently used. Noncompartmental, compartmental, and physiological models were considered in this chapter including the quantitative methodology for utilizing these models. The mechanisms of absorption and transport, metabolism, mode of toxic action, and excretion are considered elsewhere in this handbook.

REFERENCES

Ambrose, A. M., Christensen, H. E., Robbins, D. J., & Rather, L. J. (1953). Toxicological and pharmacological studies on chlordane. *Arch. Ind. Hyg. Occup. Med.*, *7*, 197–210.

Anadon, A., Martinez-Larranaga, M. R., Diaz, M. J., & Bringas, P. (1991). Toxicokinetics of permethrin in the rat. *Toxicol. Appl. Pharmacol.*, *110*, 1–8.

Arms, A. D., & Travis, C. C. (1988). *Reference physiological parameters in pharmacokinetic modeling*. Rep. NTIS PB 88-196019. Office of Health and Environmental Assessment, U.S. Environmental Protection Agency, Washington, DC.

Burgaz, S., Afkham, B. L., & Karakaya, A. E. (1994). Organochlorine pesticide contaminants in human adipose tissue collected in Ankara (Turkey) 1991–1992. *Bull. Environ. Contam. Toxicol.*, *53*, 501–508.

Casey, P. H., & Collie, W. R. (1984). Severe mental retardation and multiple congenital anomalies of uncertain cause after extreme parental exposure to 2,4-D. *J. Pediatr.*, *104*, 313–315.

Chui, Y. C., Poon, G., & Law, F. (1988). Toxicokinetics and bioavailability of paraquat in rats following different routes of administration. *Toxicol. Ind. Health*, *4*, 203–219.

Dale, W. E., Gaines, T. B., & Hayes, W. J., Jr. (1962). Storage and excretion of DDT in starved rats. *Toxicol. Appl. Pharmacol.*, *4*, 89–106.

Davidow, B., & Frawley, J. P. (1951). Tissue distribution, accumulation and elimination of the isomers of benzene hexachloride. *Proc. Soc. Exp. Biol. Med.*, *76*, 780–783.

DeJongh, J., & Blaauboer, B. J. (1997). Simulation of lindane kinetics in rats. *Toxicology*, *122*, 1–9.

Durham, W. F. (1969). Body burden of pesticides in man. *Ann. N.Y. Acad. Sci.*, *160*, 183–195.

el-Masri, H. A., Thomas, R. S., Benjamin, S. A., & Yang, R. S. H. (1995). Physiologically based pharmaco-kinetic/pharmacodynamic modeling of chemical mixtures and possible applications in risk assessment. *Toxicology, 105,* 275–282.

el-Masri, H. A., Thomas, R. S., Sabados, G. R., Phillips, J. K., Constan, A. A., Benjamin, S. A., et al. (1996). Physiologically based pharmacokinetic/pharmacodynamic modeling of the toxicologic interaction between carbon tetrachloride and Kepone. *Arch. Toxicol., 70,* 704–713.

Fisher, H. L., Hall, L. L., Sumler, M. R., & Shah, P. V. (1992). Dermal penetration of [14C]captan in young and adult rats. *J. Toxicol. Environ. Health, 36,* 251–271.

Freeman, R. A., Rozman, K. K., & Wilson, A. G. E. (1989). Physiological pharmacokinetic model of hexa-chlorobenzene in the rat. *Health Phys., 57*(Suppl. 1), 139–147.

Garcia-Repetto, R., Martinez, D., & Repetto, M. (1995). Coefficient of distribution of some organophos-phorous pesticides in rat tissue. *Vet. Hum. Toxicol., 37,* 226–229.

Gearhart, J. M., Jepson, G. W., Clewell, H. J., Andersen, M. E., & Conolly, R. B. (1994). Physiologically based pharmacokinetic model for the inhibition of acetylcholinesterase by organophosphate esters. *Environ. Health Perspect., 102*(Suppl. 11), 51–60.

Gibaldi, M., & Perrier, D. (1982). *Pharmacokinetics* (2nd ed.). New York: Dekker.

Hayes, W. J., Jr., Quinby, G. E., Walker, K. C., Elliott, J. W., & Upholt, W. M. (1958). Storage of DDT and DDE in people with different degrees of exposure to DDT. *Arch. Ind. Health, 18,* 398–406.

Hissink, A. M., Van Ommen, B., Kruse, J., & Van Bladeren, P. J. (1997). A physiological based pharmaco-kinetic (PB-PK) model for 1,2-dichlorobenzene linked to two possible parameters of toxicology. *Toxicol. Appl. Pharmacol., 145,* 301–310.

Kim, C. S., Gargas, M. L., & Andersen, M. E. (1994). Pharmacokinetic modeling of 2,4-dichlorophenoxy-acetic acid (2,4-D) in rat and in rabbit brain following single dose administration. *Toxicol. Lett., 74,* 189–201.

Kim, C. S., Slikker, W., Jr., Binienda, Z., Gargas, M. L., & Andersen, M. E. (1995). Development of a physi-ologically based pharmacokinetic model for 2,4-dichlorophenoxyacetic acid dosimetry in discrete areas of the rabbit brain. *Neurotoxicol. Teratol., 17,* 111–120.

Kim, C. S., Binienda, Z., & Sandberg, J. A. (1996). Construction of a physiologically based pharmaco-kinetic model for 2,4-dichlorophenoxyacetic acid dosimetry in the developing rabbit brain. *Toxicol. Appl. Pharmacol., 136,* 250–259.

Krishnan, K., & Andersen, M. E. (1994). Physiologically based pharmacokinetic modeling in toxicology. In A. W. Hayes (Ed.), *Principles and methods of toxicology* (3rd ed., pp. 149–188) New York: Raven Press.

Lehman, A. J. (1956). The minute residue problem. *Q. Bull. Assoc. Food Drug Off., 20,* 95–99.

Leung, H. W., & Paustenbach, D. J. (1988). Application of pharmacokinetics to derive biological exposure indexes from threshold limit values. *Am. Ind. Hyg. Assoc. J., 49,* 445–450.

Ludwig, G., Weis, J., & Korte, F. (1964). Excretion and distribution of aldrin-14C and its metabolites after oral administration for a long period of time. *Life Sci., 3,* 123–130.

Robinson, J., Roberts, M., Baldwin, M., & Walker, A. I. T. (1969). The pharmacokinetics of HEOD (diel-drin) in the rat. *Food Cosmet. Toxicol., 7,* 317–332.

Roth, W. L., Freeman, R. A., & Wilson, A. G. E. (1993). A physiologically based model for gastrointestinal absorption and excretion of chemicals carried by lipids. *Risk Anal., 13,* 531–543.

Sandberg, J. A., Duhart, H. M., Lipe, G., Binienda, Z., Slikker, W., Jr., & Kim, C. S. (1996). Distribution of 2,4-dichlorophenoxyacetic acid (2,4-D) in maternal and fetal rabbits. *J. Toxicol. Environ. Health, 49,* 497–509.

Street, J. C. (1964). DDT antagonism to dieldrin storage in adipose tissue of rats. *Science, 146,* 1580–1581.

Thies, M. L., & McBee, K. (1994). Cross-placental transfer of organochlorine pesticides in Mexican free-tailed bats from Oklahoma and New Mexico. *Arch. Environ Contam. Toxicol., 27,* 239–242.

Timchalk, C. (2004). Comparative inter-species pharmacokinetics of phenoxyacetic acid herbicides and related organic acids: Evidence that the dog is not a relevant species for evaluation of human health risk. *Toxicology, 200,* 1–19.

Timchalk, C., Dryzga, M. D., Johnson, K. A., Eddy, S. L., Freshour, N. L., & Kropscott, B. E., et al. (1996). Comparative pharmacokinetics of [14C]metosulam (N-[2,6-dichloro-3-methylphenyl]-5,7-dimethoxy-1,2,4-triazolo[1,5a]-pyrimidine-2-sulfonamide) in rats, mice and dogs. *J. Appl. Toxicol., 17,* 9–21.

Timchalk, C., Nolan, R. J., Mendrala, A. L., Dittenber, D. A., Brsak, K. A., & Mattsson, J. L. (2002). A physiologically based pharmacokinetic and pharmacodynamic (PBPK/PD) model for the organophosphate insecticide chlorpyriphos in rats and humans. *Toxicol. Sci.*, *66*, 34–53.

Toutain, P. L., & Bousquet-Melou, A. (2004a). Volumes of distribution. *J. Vet. Pharmacol. Ther.*, *27*, 441–453.

Toutain, P. L., & Bousquet-Melou, A. (2004b). Plasma terminal half-life. *J. Vet. Pharmacol. Ther.*, *27*, 427–439.

Wagner, J. G. (1967). Drug accumulation. *J. Clin. Pharmacol.*, *7*, 84–88.

Wagner, J. G., Northam, J. I., Alway, C. D., & Carpenter, O. S. (1965). Blood levels of drug at the equilibrium state after multiple dosing. *Nature*, *207*, 1301–1302.

Woollen, B. H. (1993). Biological monitoring for pesticide absorption. *Ann. Occup. Hyg.*, *37*, 525–540.

Wu, H. X., Evreux-Gros, C., & Descotes, J. (1996). Diazinon toxicokinetics, tissue distribution and anticholinesterase activity in the rat. *Biomed. Environ. Sci.*, *9*, 359–369.

Yang, R. S., el-Masri, H. A., Thomas, R. S., & Constan, A. A. (1995). The use of physiologically-based pharmacokinetic/pharmacodynamic dosimetry models for chemical mixtures. *Toxicol. Lett.*, *82–83*, 497–504.

Yang, R. S., el-Masri, H. A., Thomas, R. S., Constan, A. A., & Tessari, J. D. (1995). The application of physiologically based pharmacokinetic/pharmacodynamic (PBPK/PD) modeling for exploring risk assessment approaches of chemical mixtures. *Toxicol. Lett.*, *79*, 193–200.

Zatz, J. L. (1972). Accumulation of organochlorine pesticides in man. *J. Pharm. Sci.*, *61*, 948–949.

CHAPTER 7

Metabolic Interactions of Pesticides

Ernest Hodgson
North Carolina State University, Raleigh, NC, USA

Outline

CHEMICAL FACTORS AFFECTING PESTICIDE METABOLISM: INTRODUCTION

Although the study of the metabolism and toxicity of pesticides is simplified by considering single compounds, humans and other living organisms are not exposed in this way; rather, they are exposed to many xenobiotics simultaneously, involving different portals of entry, modes of action, and metabolic pathways. Because they bear directly on the problem of toxicity-related interactions between various xenobiotics, metabolic interactions between exogenous compounds are important in the study of toxicity. This is particularly true in the case of pesticides, which are frequently formulated, and used, as mixtures. Also of importance in considerations of pesticide toxicity and safety are metabolic interactions between pesticides and endogenous metabolites.

Pesticide Biotransformation and Disposition
DOI: 10.1016/B978-0-12-385481-0.00007-1

Table 7.1 In Vivo Assessment of Altered Microsomal Activities in Humans and Animals Using Test Compounds

Test compound	Test	Species	Reference
Aminopyrene	Breath test	Human	Jager et al. (1980)
Antipyrine	Plasma half-life, urinary excretion	Human, rat	Mehta et al. (1982); Butler and Dauterman (1989)
Caffeine	Plasma half-life, urinary excretion	Human	Kadlubar et al. (1992); Relling et al. (1992)
Chloramphenicol	Plasma half-life	Human	Mehta et al. (1975)
Hexobarbital	Sleep time	Rat	Butler and Dauterman (1988)
β-Methyldigoxin	Plasma half-life, urinary excretion	Human	Hinderling and Garrett (1977)
Phenylbutazone	Plasma half-life	Human	Krishnaswamy et al. (1981)
Procaine	Paralysis	Rat	Butler and Dauterman (1988)
Salicylates	Plasma half-life, urinary excretion	Rat	Yu and Varma (1982)
Theophylline	Plasma half-life	Human, rat	Mehta et al. (1982); Butler and Dauterman (1988)

Pesticides, and other xenobiotics, in addition to serving as substrates for a number of xenobiotic-metabolizing enzymes (XMEs), may also serve as inhibitors or inducers of these or other enzymes. Moreover, there are many compounds that first inhibit and subsequently induce such enzymes as the microsomal monooxygenases. The situation is even further complicated by the fact that, although some substances have an inherent toxicity and are detoxified in the body, others without inherent toxicity can be metabolically activated to become potent toxicants. The following is illustrative of the situations that might occur involving two compounds: compound A, without inherent toxicity, is metabolized to a potent toxicant. In the presence of an inhibitor of its metabolism, there would be a reduction in toxic effect, while after exposure to an inducer of the activating enzymes, there would be an increase. Conversely, the toxicity of compound B, a toxicant that is metabolically detoxified, would be increased in the presence of an inhibitor and decreased in the presence of an inducer.

In addition to these possible cases, the toxicity of the inhibitor or inducer, as well as the time dependence of the effect, must also be considered, because, as mentioned previously, many xenobiotics that are initially enzyme inhibitors ultimately become inducers. Interactions between components in mixtures that are more complex than binary mixtures represent a particularly intractable program that has not yet been well resolved.

Other xenobiotics, such as clinical or other drugs, or occupational chemicals, by causing enzyme induction or inhibition, can affect the metabolism and thus the toxicity of pesticides. Conversely, pesticides, by acting as either enzyme inducers or inhibitors, can affect the metabolism of other xenobiotics, such as drugs, as well as the metabolism of endogenous compounds, such as steroid hormones. In the following sections, the discussion and examples will serve to illustrate these various interactions.

Although the mechanisms of enzyme inhibition and induction are investigated by a variety of biochemical and molecular biological techniques, it is important, for consideration of the implications of these phenomena in human health risk assessment, to demonstrate them in vivo. Some examples of methods and chemicals used for this purpose are shown in Table 7.1.

INDUCTION

Induction of Microsomal Enzyme Activity

The stimulatory effect of xenobiotics on liver microsomal enzymes was first reported in the 1950s (Brown et al., 1954; Conney et al., 1957; Miller et al., 1954; Remmer, 1958) and since then has been extensively investigated. Numerous early experiments with laboratory rodents confirmed hepatic enzyme induction, although until recently methods were not available for identification of individual isoforms and the inducer was often classified as a phenobarbital-, a 3-methyl-cholanthrene-, or a mixed-type inducer. Reviews in this area include those of Conney (1967), Gelboin and Conney (1968), Sher (1971), Gillette et al. (1972), Nebert and Jensen (1979), Okey et al. (1986), Okey (1990), Batt et al. (1992), and Denison and Whitlock (1995). Reviews with emphasis on pesticides include those of Fouts (1963), Conney et al. (1967), Leibman (1968), Street et al. (1969), DuBois (1969), Hodgson (1974), Hodgson and Kulkarni (1974), Hodgson et al. (1980), Wilkinson and Denison (1982), Khan (1984), Kulkarni and Hodgson (1984a,b), Hodgson and Levi (1996), and Hodgson and Meyer (1997, 2010). It should be noted that induction is not restricted to xenobiotics, and enzymes may also be induced by hormones and other normal body constituents (Conney, 1967; Conney et al., 1967, 1979; Kobliakov et al., 1991; Pantuck et al., 1979, 1984; Ronis and Cunny, 1994; Schenkman et al., 1989) and by dietary constituents (Anderson and Kappas, 1991; Donaldson, 1994; Hodgson and Meyer, 2010; Wattenberg, 1971).

More recently, human hepatocytes have been used as a model system for investigating the induction of microsomal monooxygenases and other XMEs by pesticides (e.g., Das et al., 2006, 2008a,b).

A large number of studies have provided evidence for induction of enzymes in surrogate animals, or in humans who have been exposed occupationally or environmentally to pesticides (Table 7.2). For the most part these studies employed noninvasive in vivo techniques such as examination of the half-life of aminopyrene or phenylbutazone or excretion of 6β-hydroxycortisol (Guzelian et al., 1980; Kolmodin et al., 1969; Kolmodin-Hedman, 1973; Kreiss et al., 1981; Poland et al., 1970). More recently, in vitro techniques have been used following in vivo exposure, techniques that yield information on such aspects as isoform specificity and the mechanism of induction.

Table 7.2 Induction of Microsomal Enzyme Activity Following Treatment, in Vivo, and Involving Pesticides as Either Inducers or Substrates

Inducer and/or substrate	Species	Effect	Reference
SYNERGISTS			
Sesoxane	Mouse	Hexobarbital sleeping time increased	Fine and Molloy (1964)
Piperonyl butoxide	Mouse	Hexobarbital sleeping time increased	Fine and Molloy (1964)
		Hexobarbital sleeping time increased up to 12 h, decreased after 24–72 h; parathion toxicity increased after 1 h, decreased after 48 h	Kamienski and Murphy (1971)
		Microsomal CYP content decreased after 2–12 h, increased after 12–36 h	Philpot and Hodgson (1971–1972)
		Induction of CYP 2B10, 1A1, and 1A2; 1A2 by an Ah-independent mechanism	Philpot and Hodgson (1971–1972); Lewandowski et al. (1990); Adams et al. (1993a,b, 1995); Ryu et al. (1996); Cook and Hodgson (1985, 1986)
CHLORINATED HYDROCARBON INSECTICIDES			
BHC	Rat	In vitro metabolism of hexobarbital increased, hexobarbital sleeping time decreased, scillicocide toxicity decreased; all isomers similar	Koransky et al. (1964)
Trichloro–237	Rat	Hexobarbital sleeping time decreased	Hart and Fouts (1963)
g-Chlordane	Rat	Hexobarbital sleeping time decreased	Hart and Fouts (1963)
Endrin	Rat	Hexobarbital sleeping time decreased	Hart and Fouts (1963)
DDT	Rat	No effect on hexobarbital sleep time	Hart and Fouts (1963)
		Hexobarbital sleep time decreased; hexobarbital metabolism in vitro increased; metabolism of aminopyrine increased; metabolism of p-nitrobenzoic acid increased; no effect on aniline metabolism; increased detoxication of EPN; increased O–demethylation of p-nitroanisole.	Hart and Fouts (1963); Kinoshita et al. (1966)
		Increase in CYP2B and CYP3A	Li et al. (1995)
DDT	Squirrel monkey	Increased N-demethylation of aminopyrine	Cranmer et al. (1972)

Compound	Species	Effect	Reference
DDT	Human	Increased metabolism, in vitro, of EPN and p-nitroanisole; increase in CNS arousal with phenobarbital administration	Rappolt (1973)
DDT and analogs	Mouse	Increased P450 levels, aniline hydroxylase activity, zoxazolamine paralysis time, and hexobarbital sleep time; QSAR for 28 analogs	Abernathy et al. (1971a,b)
Diphenyl hydantoin	Rat	DDT and DDE storage decreased, in vitro DDT metabolism increased	Cranmer (1970)
o,p-DDD	Guinea pig	Phenobarbital sleep time decreased, in vitro phenobarbital metabolism increased	Straw et al. (1965)
p,p-DDD	Rat	Increased metabolism in vitro of estradiol-17β	Welch et al. (1971)
o,p-DDT	Rat	PXR– and CAR–dependent increase in CYPs 2B2 and 3A2	Kiyosawa et al. (2008)
o,p-DDT	HepG2 cells	Increase in CYP3A4 mRNA	Medina-Diaz and Elizondo (2005); Medina-Diaz et al. (2007)
p,p-DDE	Rat	Increased metabolism in vitro of estradiol-17β	Welch et al. (1971)
		Induction of CYP 2b 3A	Wyde et al. (2003)
Chlordane	Rat	In vitro metabolism of hexobarbital, aminopyrine, and chlorpromazine unchanged after one dose; all increased after three doses	Hart and Fouts (1963)
Chlordane	Dog	Decreased toxicity of dicoumarol	Welch and Harrison (1966)
Lindane	Rat	Increased metabolism in vitro of estrone	Welch et al. (1971)
		Increased metabolism in vitro of estradiol-17β	Welch et al. (1971)
		Increased expression of mRNA for CYP1A1, 1A2, 2B1, 2B2, and 2E1 as well as associated catalytic activities	Johri et al. (2003, 2008)
Heptachlor	Rat	Increased metabolism in vitro of estradiol-17β	Welch et al. (1971)
Toxaphene	Rat	Increased metabolism in vitro of estradiol-17β	Welch et al. (1971)
		Increased detoxication of EPN, O-demethylation of p-nitroanisole and N-demethylation of aminopyrine	Kinoshita et al. (1966)
Dieldrin	Rhesus monkey	Increased metabolism in vitro of chlorfenvinphos	Wright et al. (1972)

(Continued)

Table 7.2 (Continued)

Inducer and/or substrate	Species	Effect	Reference
Dieldrin	Dog	Increased metabolism in vitro of chlorfenvinphos	Wright et al. (1972)
Dieldrin	Rat	Increased metabolism in vitro of chlorfenvinphos	Wright et al. (1972)
Dieldrin	Rat	Increased metabolism in vitro of estradiol–17β	Welch et al. (1971)
Dieldrin	Mouse	Increased metabolism in vitro of chlorfenvinphos	Wright et al. (1972)
Mirex	Mouse/rat	Increased O-demethylation in vitro of p-nitroanisole and CYP content	Baker et al. (1972)
Mirex and kepone	Rat	Increased warfarin hydroxylation in vitro and increased CYP content	Kaminsky et al. (1978)
Mirex and kepone	Gerbil	Increased benzo(a)pyrene hydroxylase activity in vitro and increased CYP content	Crouch and Ebel (1987)
Mirex and kepone	Mouse	Increased acute in vivo hepatotoxicity; mirex had greater effect	Fouse and Hodgson (1987)
		Increased CYP and N- and O-dealkylation in vitro Induction of CYP2B10, 1A2, and 3A	Fabacher and Hodgson (1976) Lewandowski et al. (1989)
Methoxychlor	Rat	Increase in total CYP and gender-dependent increases in CYP 1A1- and 1A2-related activities	Orepeza-Hernandez et al. (2003)
		Increase in CY2PB and CYP3A	Li et al. (1995)
ORGANOPHOSPHORUS INSECTICIDES			
3–Methylcholanthrene	Rat	Increased metabolism in vitro of azinphosmethyl to a cholinesterase inhibitor	Murphy and DuBois (1957)
Phenobarbital	Rat/mouse	Decreased in vivo toxicity of parathion, methyl parathion, demeton, disulfoton, azinphosmethyl, dioxathion, ethion, carbophenothion, mevinphos, and EPN	Dubois (1969)
Malathion	Rat	Increased CYP	Matthews and Devi (1994)
S,S,S–tri–n–butyl phosphorotrithioate	Hen	Increased CYP	Lapadula et al. (1984)
Chlorpyrifos	Human	Increased CYP isoforms in hepatocytes	Das et al. (2008a,b)
CARBAMATE INSECTICIDE			
Carbaryl	Chicken	Pentobarbital sleep time decreased	Puryear and Paulson (1972)

PYRETHROID INSECTICIDES			
Permethrin	Rat, human	Increased CYP isoform protein and mRNA in hepatocytes	Heder et al. (2001); Das et al. (2008a,b)
Deltamethrin	Rat, human	Increase in CYP mRNA and protein	Johri et al. (2006, 2008); Das et al. (2008a,b)
Pyrethrins	Rat	Increase in CYP2B1 and CYP2B1/2 mRNA and associated enzymatic activities; also increase in testosterone 6β-hydroxylase activity	Price et al. (2008)
Pyrethrins	Human	Increase in testosterone 6β-hydroxylase activity, CYP2B6 and CYP3A4 mRNA	Price et al. (2008)
RODENTICIDES			
Phenobarbital	Human	Pharmacological activity of warfarin decreased	Robinson and MacDonald (1966)
Phenobarbital	Dog	Dicoumarol toxicity in vivo decreased	Welch et al. (1967)
Acetylsalicylic acid	Rat	Dicoumarol prothrombin time decreased	Coldwell and Zawidzka (1968)
Heptobarbital	Human	Excretion of dicoumarol metabolites increased	Aggeler and O'Reilly (1969)
HERBICIDES			
Alachlor	Rat	Induction of CYP2B1/2 and 1A1/2 protein and associated activities	Hanioka et al. (2002)
Atrazine	Rat	Increased xenobiotic metabolism	Ugazio et al. (1993)
Metolachlor	Rat	Induction of CYP2B1/2 and CYP3A1/2 protein	Dalton et al. (2003)
Monuron 1967	Rat	Detoxication of EPN, O-demethylation of p-nitroanisole and N-demethylation of aminopyrine increase for 1–3 weeks, then return to normal	Kinoshita and DuBois (1967)
Diuron	Rat	Detoxication of EPN, O-demethylation of p-nitroanisole and N-demethylation of aminopyrine increase for 1–3 weeks, then return to normal	Kinoshita and DuBois (1967)
Diuron	Mouse hepatoma cells	AhR–dependent induction of CYP1A1 mRNA	Zhao et al. (2006)
Tridiphane	Mouse	Induction of CYP4A	Levi et al. (1992)
		Induction of epoxide hydrolase	Moody and Hammock (1987)

(Continued)

Table 7.2 (Continued)

Inducer and/or substrate	Species	Effect	Reference
FUNGICIDES			
Griseofulvin	Human	Pharmacological action of warfarin decreased	Cullen and Catalano (1967)
Parnon	Rat	In vitro metabolism increased	Hoffman et al. (1968)
Azoles	Rat	Increase in liver weight, liver pathology, total CYPs, CYPs 2B1 and 3A2, and associated enzyme activities	Barton et al. (2006); Sun et al. (2006, 2007); Martin et al. (2007)
Azoles	Mouse	Increase in liver weight, liver pathology, total CYPs, CYPs 2B1 and 3A2, and associated enzyme activities	Allen et al. (2006); Ward et al. (2006); Sun et al. (2006)
PYRAZOLE INSECTICIDE			
Fipronil	Human	Increased CYP isoforms in hepatocytes	Das et al. (2006)
PESTICIDE ADJUVANT			
Toximul	Mouse	PPAR–dependent increases in peroxisomal acyl–CoA oxidase, thiolase, and Cyp4a10 and 4a14	Upham et al. (2007)
INSECT REPELLENT			
DEET	Human	Increased CYP isoforms in hepatocytes	Das et al. (2008a,b)

Adapted from Hodgson and Meyer (2009).

Dalton et al. (2003), Das et al. (2006, 2008a,b), and Johri et al. (2006, 2008) are among the many examples cited throughout this chapter.

Abernathy et al. (1971a,b) demonstrated significant decreases in zoxazolamine paralysis time, hexobarbital sleeping time, and aniline hydroxylase activity in mice following treatment with dichlorodiphenyltrichloroethane (DDT) or 1,1-dichloro-2,2-bis(*p*-chlorophenyl)ethylene (DDE), a major metabolite of DDT and a persistent residue in animals, including humans, even in countries where DDT use has been banned for decades.

Different inducers may increase the expression of different enzymes and, therefore, different metabolic pathways. Thus, Chadwick et al. (1971) showed that repeated doses of lindane or DDT increased oxidative hydrolysis, O-demethylase, dehydrochlorinase, and glucuronyl transferase activity, but to different degrees. Pretreatment of rats with lindane caused them to metabolize a single dose of radioactive lindane 2.5 times more extensively than controls, and pretreatment with DDT caused a 3.5-fold increase in metabolism of radioactive lindane. Furthermore, the DDT pretreatment was followed by generation of proportionally more neutral and weakly polar, but less free-acid-type, metabolites of the radioactive lindane. Thus, metabolism was qualitatively as well as quantitatively different following administration of the two inducers. Subsequent studies (Chadwick and Freal, 1972) confirmed these findings, including the increased excretion of metabolites following pretreatment with DDT. In addition, it was shown that rats pretreated with DDT plus lindane excreted more 2,4,5-trichlorophenol and 2,3,4,6- and 2,3,4,5-tetrachlorophenols by the second day of treatment than did rats receiving lindane alone. The results suggested that DDT treatment stimulates the metabolism of lindane through a selective effect on certain metabolic pathways involved in its oxidative degradation, notably those leading to the formation of tetrachlorophenols, particularly 2,3,4,5-tetrachlorophenol.

Incidentally, when two inducers are involved, the resulting induction may be either additive or slightly antagonistic. Thus, Gielen and Nebert (1971) found an additive effect when either phenobarbital or *p,p'*-DDT was present with a polycyclic hydrocarbon, but not when combinations of phenobarbital plus DDT or one polycyclic hydrocarbon plus another were involved.

Although it had been known for many years that various pesticides could induce cytochrome P450s (CYPs), neither the specific isozymes induced nor the implications of the induction were well characterized. Later studies using enzymatic and immunochemical techniques examined isoform specificity of specific pesticides. For example, mirex and chlordecone were shown to induce CYP2B10 and testosterone metabolism in mouse liver, a pattern of induction similar to that of phenobarbital (Baker et al., 1972; Fabacher and Hodgson, 1976; Hodgson, 1974; Lewandowski et al., 1989). Enzymatic activities suggested that, in addition to 2B10, other CYPs were induced, and later studies demonstrated induction of CYP1A2 and CYP3A (Dai et al., 1998; Hodgson and Levi, 1996).

Another group of pesticides, the phenoxyacetic acid herbicides (e.g., 2,4-dichloro-phenoxyacetic acid), and the herbicide synergist tridiphane were found to induce the CYP4A isozymes in rodents (Levi et al., 1992; Moody et al., 1987, 1992). These CYP isoforms are known to be involved in the oxidation of fatty acids and the mainte-nance of lipid homeostasis. Moreover, in rodents, compounds that are CYP4A inducers also cause peroxisome proliferation, an event associated with nongenotoxic induction of liver tumors in rodents. Another peroxisome proliferator, the fenvalerate metabolite fenvaleric acid, has been shown to induce several CYP-dependent enzyme activities including 7-ethoxyresorufin de-ethylation, catalyzed by CYP1As; 7-pentoxyresorufin O-dealkylation, catalyzed by CYP2Bs; and testosterone hydroxylation, catalyzed by CYP3A and CYP2B11 (Morisseau et al., 1991). The exact relationship of these inter-actions and the relevance to humans has not yet been defined.

Methylenedioxyphenyl (MDP) compounds, such as piperonyl butoxide and sesa-mex, have been used as synergists with pyrethroid and carbamate pesticides. Other well-known MDP compounds such as safrole and isosafrole are found in many com-mon foods of plant origin. These chemicals affect multiple enzyme pathways, including the CYP system (Hodgson and Philpot, 1974). Their effect on CYP enzymes is bipha-sic, that is, inhibition followed by induction, and is discussed more fully under Biphasic Effects: Inhibition and Induction. Reviews include those of Philpot and Hodgson (1971–1972), Hodgson and Levi (1998), and Hodgson (1999). However, MDP com-pounds are known to induce both enzyme and mRNA for several CYP isoforms in the mouse, including CYP1A1, CYP1A2, and CYP2B10. CYP1A2 is induced by both Ah-receptor-dependent and Ah-receptor-independent mechanisms (Cook and Hodgson, 1985, 1986; Ryu et al., 1995, 1996).

The fungicide captan, although inhibiting many hepatic CYP-dependent activities in mouse liver (Paolini et al., 1999), induces both CYP3A and CYP1A2 in the kidney and CYP1A2 in the lung. The ergosterol biosynthesis inhibiting fungicides (EBIFs), for example, clotrimazole and propioconazole, have been shown to have multiple effects on the rodent CYP system (Ronis et al., 1994). The EBIFs induced CYPs 3A, 2B, and 1A, while suppressing the activity of CYP2C11. These alterations were found to cause significant changes in testosterone metabolism in male rats. Cellular techniques are becoming more available for the study of induction by pesticides.

The azole fungicides have been shown to be inducers of various XMEs, primarily in rodents (Allen et al., 2006; Barton et al., 2006; Martin et al., 2007; Sun et al., 2006, 2007; Ward et al., 2006).

Among recent studies of interest is the demonstration of the induction of CYP4A10 and CYP4A14 in mice by the pesticide adjuvant, Toximul, an effect medi-ated through the PPARα receptor (Upham et al., 2007). Diuron and related phenyl-urea herbicides induced CYP1A1 via the Ah receptor in several cell lines, including at least one human cell line (Zhao et al., 2006). Further studies showed the induction, in rodents, of CYPs 2B and 3A by the herbicide metolachlor (Dalton et al., 2003), of CYPs 1A and 2B by deltamethrin (Johri et al., 2006), and of a number of CYP-related metabolic activities by the herbicide alachlor (Hanioka et al., 2002).

DuBois et al. (1996) used hepatocytes from the rat and quail as well as human hepatoma-derived (HepG2) cells to study the induction of CYP isoforms by pesticides. The pesticides fell into four groups: first, CYP3A inducers such as pentachlorophenol; second, 3-methylcholanthrene-type inducers, such as lindane, an inducer of CYP1A isoforms; third, phenobarbital-type inducers, such as dieldrin, an inducer of CYP2B isoforms; and fourth, pesticides with little or no capacity to induce CYP isoforms. Pentachlorophenol and lindane were the strongest inducers in these cell lines, and lindane appeared to be a member of both the second and the third groups because it induced both CYP1A and CYP2B activities.

Although a small number of studies showed induction by pesticides in rat hepatocytes, for example, the induction of CYP2B1 by pyrethroids (Heder et al., 2001), the recent availability of human hepatocytes has enabled the investigation of induction by pesticides in humans (Das et al., 2006, 2008a,b). These studies reveal that fipronil is an effective inducer of CYP1A1, CYP2B6, and CYP3A4 in human hepatocytes, at concentrations as low as 1.0 μM. The pyrethroids deltamethrin and permethrin, while not as effective as fipronil, induced the same CYP isoforms. The insecticide chlorpyrifos and the insect repellent DEET are also capable of inducing xenobiotic-metabolizing CYP isoforms in human hepatocytes. All of the results of studies utilizing human hepatocytes suggest that pesticide-pesticide interactions are possible in the human liver, and this possibility should be investigated. It might also be noted that, to a greater or lesser extent, these pesticides are also cytotoxic to human hepatocytes but generally at higher concentrations than those required for CYP isoform induction.

Of particular concern is the ability of many pesticides to disrupt the normal functioning of the endocrine system, and endocrine disruption has become an important environmental concern (Birnbaum, 1994; Colborn et al., 1993; Guillette et al., 1994). For this reason, many of the recent studies of induction by organochlorines have involved either methoxychlor or o,p-DDT. Methoxychlor has been shown to induce CYPs 1A, 2B, 2C, 2E, and 3A in both male and female rats (Oropeza-Hernandez et al., 2003) and to induce ethoxyresorufin O-deethylase and pentoxyresorufin O-depentylase activity in HepG2 cells (Dehn et al., 2005; Medina-Diaz and Elizondo, 2005). o,p-DDT and its metabolite o,p-DDE have been shown to induce CYP3A4 (Medina-Diaz et al., 2007). The finding that the nuclear receptors pregnane X receptor (PXR) and constitutive androstane receptor (CAR) are involved in induction in both HepG2 cells and in immature ovariectomized rats (Kiyosawa et al., 2008; Wyde et al., 2003) is of mechanistic importance. Lindane has been shown to induce CYPs 1A and 2B in rats (Johri et al., 2008; Parmar et al., 2003).

It is now quite clear that pesticides may have complex effects on the hepatic monooxygenase system and, in addition to affecting xenobiotic metabolism, pesticides, by disturbing endogenous metabolism, have the potential to produce profound changes in both the physiological and the reproductive capacities of the organism (Birnbaum, 1994; Tyler et al., 1998).

Induction of Other Enzymes

Microsomal enzymes are not the only ones subject to induction. For example, δ-aminolevulinic acid (ALA) synthetase is located in the mitochondria and its activity may increase 40–100 times in those structures on induction (Granick, 1965).

Interaction between the induction of mitochondrial and that of microsomal enzymes is illustrated by the action of the pesticide *m*-dichlorobenzene in rats. Following daily doses at the rather high rate of 800 mg/kg, there is a biphasic stimulation of ALA-synthetase activity and of the excretion of urinary coproporphyrin, both of which peak by 3 days and then decline. The decrease in ALA-synthetase and in excretion of coproporphyrin at 5 days corresponds to the maximal stimulation of drug metabolism and to a decrease in the concentration of *m*-dichlorobenzene in the serum and liver at the same time (Poland et al., 1971).

In the cytosolic fraction of homogenized rat liver, the activity of nicotinamide adenine dinucleotide (NAD)-dependent aldehyde dehydrogenase (EC 1.2.1.3) is increased up to 10-fold after administration of phenobarbital for 3 days. The effect is genetically controlled and is inherited as an autosomal dominant characteristic. The mechanism is apparently unrelated to other drug-induced increases in enzyme activity, such as those that occur in the hepatic microsomal systems for drug metabolism (Deitrich, 1971). Glutathione *S*-transferases as well as CYPs were shown to be induced by pesticides, but the levels of induction of the former were much lower (Fabacher et al., 1980; Hodgson et al., 1980; Kulkarni et al., 1980; Robacker et al., 1981).

Mechanism of Induction

Several pesticides have been tested for activation of the PXR receptor in engineered expression systems (Lemaire et al., 2006; Matsubara et al., 2007). Since human 3A4 is transcriptionally regulated, in part, by PXR, induction of this CYP is expected for the pesticides active in the in vitro systems. In rats, one of these compounds, the herbicide metolachlor, was shown to induce the rat ortholog CYP3A2 as well as CYPB1/2 at approximately one-fifth the potency of phenobarbital (Dalton et al., 2003). Similarly, PXR activation by conazole fungicides is supported by activity in the PXR expression system and from toxicogenomic profile-like signatures of prototype PXR ligands (Goetz et al., 2006; Tully et al., 2006). Similar approaches have defined transcriptional mechanisms for the pesticide induction of other CYPs through their activity as ligands for relevant transcriptional activators. Casabar et al. (2010) utilized the HepG2 cell line and primary human hepatocytes to investigate the induction of CYP2B6 and CYP3A4 by endosulfan-α. They demonstrated the involvement of the human PXR and the human CAR in the inductive process. Endosulfan-α is a strong activator of PXR and inducer of CYP2B6 and CYP 3A4. CAR, on the other hand, is only weakly activated.

The Ah receptor is known to be involved in the induction of Cyp isoforms in mice by MDP and related compounds (Cook and Hodgson, 1985, 1986) and may be assumed to be involved in any induction of Cyp1a1 or CYP1A1. The PPARα receptor has been shown to be involved in the induction, in mice, of Cyp4a10 and Cyp4a14 (Upham et al., 2007).

INHIBITION

As previously indicated, inhibition of XMEs can cause either an increase or a decrease in toxicity. Inhibitory effects can be demonstrated in a number of ways at various organizational levels.

Types of Inhibition and Experimental Demonstration
In Vivo Symptoms

The measurement of the effect of an inhibitor (or inducer) on the duration of action of a drug in vivo is a common method of demonstrating its action, and previously the effects on the hexobarbital sleeping time or the zoxazolamine paralysis time were often used. Both of these drugs are fairly rapidly deactivated by the hepatic microsomal CYP-dependent monooxygenase system; thus, inhibitors of the CYP isoform(s) involve prolonging their action, whereas inducers have the opposite effect. Although, as a consequence of the availability of single, expressed isoforms for direct studies of inhibitory mechanisms, these methods are now used much less often, they are still valuable for demonstrating the effect in the intact organism, a necessity for risk assessment.

In the case of activation reactions, such as the activation of the insecticide azinphosmethyl to its potent anticholinesterase oxon derivative, a decrease in toxicity is apparent when rats are pretreated with the CYP inhibitor SKF-525A.

Distribution and Blood Levels

Treatment of an animal with an inhibitor of xenobiotic metabolism may cause changes in the blood levels of an unmetabolized toxicant and/or its metabolites. This procedure may be used in the investigation of the inhibition of detoxication pathways; it has the advantage over in vitro methods of yielding results of direct physiological or toxicological interest because it is carried out in the intact animal. Moreover, the time sequence of the effects can be followed in individual animals, a factor of importance when inhibition is followed by induction.

A refinement of this technique is to determine the effect of an inhibitor on the overall metabolism of a xenobiotic in vivo, by following the appearance of metabolites in the urine and feces and/or in blood or other tissue. Again, the use of the intact animal has practical advantages over in vitro methods, although little is revealed about the mechanisms involved.

Effects on in Vitro Metabolism Following in Vivo Treatment

This method of demonstrating inhibition is of variable utility. The preparation of enzymes from animal tissues usually involves considerable dilution with the preparative medium during homogenization, centrifugation, and resuspension. As a result, inhibitors not tightly bound to the enzyme in question are lost, either in whole or in part, during the preparative processes. Therefore, negative results can have little utility because failure to inhibit and loss of the bound inhibitor give identical results. Positive results, on the other hand, not only indicate that the compound administered is an inhibitor but also provide a clear indication of excellent binding to the enzyme, most probably due to the formation of a covalent or slowly reversible inhibitory complex. The inhibition of acetylcholinesterase following treatment of the animal with organophosphorus compounds, such as paraoxon, is a good example, because the phosphorylated enzyme is stable and is still inhibited after the preparative procedures. In contrast, inhibition by carbamates is greatly reduced by the same procedures, because the carbamylated enzyme is unstable and, in addition, the residual carbamate is diluted.

Microsomal monooxygenase inhibitors that form stable inhibitory complexes with CYPs, such as SKF-525A, piperonyl butoxide and other methylenedioxyphenyl compounds, amphetamine and its derivatives, and organophosphorus pesticides (OPs) containing the $P = S$ moiety, can be readily investigated in this way because the microsomes isolated from pretreated animals have a reduced capacity to oxidize many xenobiotics.

In Vitro Effects

In vitro measurement of the effect of one xenobiotic on the metabolism of another is by far the most common type of investigation of interactions involving inhibition. Although it is the most useful method for the study of inhibitory mechanisms, particularly when purified enzymes are used, it is of more limited utility in assessing the toxicological implications for the intact animal. The principal reason for this is that in vitro measurement of inhibition does not assess the effects of factors that affect absorption, distribution, and prior metabolism, all of which occur before the inhibitory event under consideration and affect the concentration of inhibitor at the site of action.

The primary considerations in studies of inhibition mechanisms are reversibility and selectivity. The inhibition kinetics of reversible inhibition give considerable insight into the reaction mechanisms of enzymes and, for that reason, have been well studied. In general, reversible inhibition involves no covalent binding, occurs rapidly, and can be reversed by dialysis or by dilution. Reversible inhibition is usually divided into competitive inhibition, uncompetitive inhibition, and noncompetitive inhibition. Because these types are not rigidly separated, many intermediate classes have been described. Although enzyme kinetics of microsomal enzymes are intrinsically difficult to elucidate because the enzymes are membrane bound and both substrate and inhibitor are frequently lipophilic, methods for analysis of kinetic data are available that simplify the determination of the type of inhibition. For example, see the methods used in the study of the irreversible, mechanism-based inhibition of estradiol metabolism in humans by chlorpyrifos (Usmani et al., 2006).

Competitive inhibition is usually caused by two substrates competing for the same active site. Following classical enzyme kinetics, there should be a change in the apparent K_m but not in V_{max}. In microsomal monooxygenase reactions, type I ligands, which often appear to bind as substrates but do not bind to the heme iron, might be expected to be competitive inhibitors, and this frequently appears to be the case.

Uncompetitive inhibition has seldom been reported in studies of xenobiotic metabolism. It occurs when an inhibitor interacts with an enzyme-substrate complex but cannot interact with free enzyme. Both K_m and V_{max} change by the same ratio, giving rise to a family of parallel lines in a Lineweaver-Burke plot.

Noncompetitive inhibitors can bind to both the enzyme and the enzyme-substrate complex to form either an enzyme-inhibitor complex or an enzyme-inhibitor-substrate complex. The net result is a decrease in V_{max} but no change in K_m. Metyrapone, a well-known inhibitor of monooxygenase reactions, can also, under some circumstances, stimulate metabolism in vitro. In either case, the effect is noncompetitive, in that the K_m does not change, whereas V_{max} does, decreasing in the case of inhibition and increasing in the case of stimulation.

Irreversible inhibition, which is much more important toxicologically, can arise from various causes. In most cases, the formation of covalent or other stable bonds or the disruption of the enzyme structure is involved. In these cases, the effect cannot be readily reversed in vitro by either dialysis or dilution. The formation of stable inhibitory complexes may involve the prior formation of a reactive intermediate that then interacts with the enzyme ("suicide" or mechanism-based inhibition). An excellent example of this type of inhibition is the effect of the insecticide synergist piperonyl butoxide on hepatic microsomal monooxygenase activity, reviewed by Hodgson and Levi (1998) and Hodgson (1999). This methylenedioxyphenyl compound can form a stable inhibitory complex that blocks CO binding to CYP and also prevents substrate oxidation. This complex results from the formation of a reactive intermediate, and the type of inhibition changes from competitive to irreversible as metabolism, in the presence of NADPH and oxygen, proceeds. It appears probable that the metabolite in question is a carbene formed spontaneously by elimination of water following hydroxylation of the methylene carbon by the cytochrome (Dahl and Hodgson, 1979). Piperonyl butoxide inhibits the in vitro metabolism of many substrates of the monooxygenase system, including aldrin, ethylmorphine, aniline, aminopyrene, carbaryl, biphenyl, hexobarbital, and *p*-nitroanisole.

The inhibition, by organophosphorus compounds such as ethyl *p*-nitrophenol thiobenzene phosphonate (EPN), of the carboxylesterase that hydrolyzes malathion is a further example of xenobiotic interaction resulting from irreversible inhibition, because in this case the enzyme is phosphorylated by the inhibitor. Oxons, such as chlorpyrifos oxon, are potent inhibitors of the esterases, in humans, that hydrolyze pyrethroids such as permethrin (Choi et al., 2004).

Oxidative desulfuration of phosphorothioate pesticides such as chlorpyrifos, parathion, and fenitrothion by CYPs is known to release atomic sulfur, which covalently binds to and inactivates CYPs (Halpert et al., 1980; Kamataki and Neal, 1976; Levi et al., 1988; Neal and Halpert, 1982). Administration of fenitrothion at a dose as low as 7 mg/kg inhibited the hydroxylation of 17β-estradiol by hepatic microsomes (Berger and Sultatos, 1996). Usmani et al. (2003, 2006) have shown that chlorpyrifos and other OPs containing the $P = S$ moiety are potent inhibitors of the human metabolism of both testosterone and estradiol. These results suggest that, in low concentrations, organophosphorus insecticides have the potential to inhibit enzymes important in normal sexual development.

Another class of irreversible inhibitors of toxicological significance consists of those compounds that bring about the destruction of the xenobiotic-metabolizing enzymes. The drug allylisopropylacetamide, as well as other allyl compounds, has long been known to cause the breakdown of CYPs and the resultant release of heme; the hepatocarcinogen vinyl chloride has also been shown to have a similar effect, probably also mediated through the generation of a highly reactive intermediate.

Synergism and Potentiation

The terms synergism and potentiation have been variously used and defined but, in any case, involve a toxicity that is greater when two compounds are given simultaneously (or sequentially within a short time frame) than would be expected from a consideration of the toxicities of the compounds given alone.

An example of synergism has already been mentioned. Piperonyl butoxide, sesamex, and related compounds increase the toxicity of insecticides to insects by inhibiting insect CYPs. Other insecticide synergists that interact with CYP include aryloxyalkylamines such as SKF-525A, Lilly 18947, and their derivatives; compounds containing acetylenic bonds such as aryl-2-propynyl phosphate esters containing propynyl functions; phosphorothionates; benzothiadiazoles; and some imidazole derivatives. Insecticide synergists have similar interactions with mammalian CYPs.

The best known example of potentiation involving insecticides and an enzyme other than the monooxygenase system is the increase in the toxicity of malathion to mammals that is brought about by certain other organophosphates. Malathion has a low mammalian toxicity, due primarily to its rapid hydrolysis by a carboxylesterase.

EPN, another organophosphorus insecticide, causes a dramatic increase in malathion toxicity to mammals at dose levels that, given alone, cause essentially no inhibition of cholinesterase. In vitro studies have shown that the oxygen analog of EPN, as well as oxons of many other organophosphate compounds, increases the toxicity of malathion by inhibiting the carboxylesterase responsible for its degradation.

Antagonism

In toxicology, antagonism may be defined as that situation in which the toxicity of two or more compounds administered together (or sequentially within a short time frame) is less than would be expected from a consideration of their toxicities when administered individually. Apart from the effects mediated through induction of XMEs (discussed previously), antagonism does not appear to be important in pesticide interactions.

Pesticides as Inhibitors

Examples of pesticides as inhibitors of the metabolism of other pesticides, other xenobiotics, or endogenous metabolites in humans are shown in Table 7.3. Pesticides may act to inhibit CYPs or other enzymes by any of the mechanisms discussed previously—competitive inhibition, noncompetitive inhibition, or irreversible inhibition. As discussed more fully in the following section, MDP compounds have very complex interactions with the CYP system, being both inhibitors and inducers. Because MDP compounds are substrates for CYP enzymes, they may act initially as competitive inhibitors. As the MDP compound is metabolized, it becomes a suicide inhibitor with its reactive metabolite bound to the heme iron of CYP (Goldstein et al., 1973; Hodgson et al., 1998; Hodgson and Philpot, 1974).

The herbicide synergist tridiphane, a postemergent herbicide, owes its activity to its ability to inhibit glutathione S-transferases. It has also been shown to induce epoxide hydrolase and CYP, specifically CYP4A, and peroxisomal enzymes (Levi et al., 1992; Moody and Hammock, 1987). In addition to induction of CYP4A, tridiphane functions as a selective CYP inhibitor, inhibiting CYP2B10 while having little or no effect on other CYP isoforms (Moreland et al., 1989). As assessed by in vitro studies, tridiphane appears to be a competitive inhibitor of CYP; its effect in vivo, however, is not yet known.

Organophosphorus insecticides such as chlorpyrifos, parathion, and others that contain the $P = S$ moiety are metabolized by the CYP system to the corresponding oxon, $P = O$, by oxidative desulfuration. This activation reaction, which converts the relatively inactive compound to a potent cholinesterase inhibitor, is thought to involve the formation of a P–S–O (phosphooxythirane) ring intermediate. Studies with both microsomes and purified enzymes (Halpert et al., 1980; Kamataki and Neal, 1976; Morelli and Nakatsugawa, 1978) have demonstrated that, during oxidative desulfuration, the released sulfur exists as a highly reactive molecule that then binds to the heme iron of CYP, inactivating the enzyme. This binding of reactive sulfur to CYP is

Table 7.3 Inhibition of Human Hepatic Phase I Metabolism by Pesticides

Substrate	Enzyme	Inhibitor(s)	Reference
XENOBIOTIC SUBSTRATES			
Carbaryl	Liver microsomes	Chlorpyrifos	Tang et al. (2002)
Carbaryl	CYP2B6	Chlorpyrifos	Tang et al. (2002)
Carbofuran	Liver microsomes	Chlorpyrifos	Usmani et al. (2004)
DEET	Liver microsomes	Chlorpyrifos	Usmani et al. (2002)
Fipronil	Liver microsomes	Chlorpyrifos	Joo et al. (2007)
Fipronil	CYP3A4	Chlorpyrifos	Joo et al. (2007)
Imipramine	Liver microsomes	Chlorpyrifos, azinphosphos methyl, parathion	Di Consiglio (2005)
Imipramine	CYPs 1A2, 3A4, 2C19	Chlorpyrifos, azinphosphos methyl, parathion	Di Consiglio (2005)
Nonane	Liver microsomes	Chlorpyrifos	Joo et al. (2007)
Nonane	CYP2B6	Chlorpyrifos	Joo et al. (2007)
ENDOGENOUS SUBSTRATES			
Estradiol	Liver microsomes	Chlorpyrifos, fonofos, carbaryl, naphthalene	Usmani et al. (2006)
Estradiol	CYP1A2	Chlorpyrifos, fonofos, carbaryl, naphthalene	Usmani et al. (2006)
Estradiol	CYP3A4	Chlorpyrifos, fonofos, deltamethrin, permethrin	Usmani et al. (2006)
Testosterone	Liver microsomes	Methoxychlor	Li et al., 1993
Testosterone	Liver microsomes	Chlorpyrifos, phorate, fonofos	Usmani et al. (2003)
Testosterone	CYP3A4	Chlorpyrifos	Usmani et al. (2003)
Testosterone	CYP	Fipronil	Tang et al. (2004)

accompanied by loss of CYP as detected by measurement of the dithionite-reduced CO complex as well as loss of monooxygenase activity (Berger and Sultatos, 1996; Butler and Murray, 1993; Neal, 1985; Neal and Halpert, 1982; Neal et al., 1983). Cohen (1984) showed that acetaminophen toxicity was reduced by the organophosphorus insecticide fenitrothion, as a result of inhibition of the CYP-dependent activation of acetaminophen. Studies with purified CYP isoforms and fenitrothion demonstrated that the amount of inhibition varied with the CYP isoform (Levi et al., 1988). Other OPs, known to be CYP inhibitors, affect the distribution of neonicotinoids in the brain and, to a lesser extent, in the liver (Shi et al., 2009).

In human liver microsomes, metabolism of parathion resulted in a concurrent loss of total CYP as well as the loss of several CYP-mediated enzyme activities (Butler and Murray, 1997). The activities inhibited included testosterone 6β-hydroxylation, catalyzed by CYP3A4; 7-ethoxyresorufin deethylation, catalyzed by CYP1A2; and tolbutamide methyl hydroxylation, catalyzed by CYP2C9/10. Aniline 4-hydroxylation, catalyzed by CYP2E1, was not inhibited.

The inhibition of CYP-dependent monooxygenations by organophosphorus insecticides may be of considerable importance in human health risk assessment, as it has been shown that organophosphorus insecticides containing the P=S moiety are potent inhibitors of the human microsomal metabolism of both testosterone and estradiol (Usmani et al., 2003, 2006).

Organophosphorus compounds may also inhibit enzymes other than CYP, particularly esterases (e.g., Cohen, 1984; Gaughan et al., 1980). Gaughan et al. (1980) also showed that profenofos, EPN, and *S,S,S*-tributylphosphorotrithioate, when administered in vivo to mice, all inhibited the liver microsomal esterases hydrolyzing *trans*-permethrin as well as the carboxylesterase hydrolyzing malathion. Chlorpyrifos oxon is a potent inhibitor of the hydrolysis of pyrethroids by human liver preparations (Choi et al., 2004).

While malathion is the safest organophosphorus insecticide in current use, the presence of isomalathion, an occasional contaminant in commercial preparations, creates a serious toxicity situation because of the ability of isomalathion to inhibit the carboxylase that detoxifies malathion. This permits an increase in in vivo levels of malathion, which can then be metabolized, by CYP-dependent oxidative desulfuration, to the potent acetylcholinesterase inhibitor malaoxon (Jensen and Whatling, 2010).

The fungicide captan, apparently through reactive metabolites, inhibits several CYP-dependent enzyme activities in mouse liver (Paolini et al., 1999), although it induces the 2α-hydroxylation of testosterone. Methoxychlor, again through a reactive intermediate, inhibits the oxidation of both testosterone and estradiol, the pattern of metabolites indicating inhibition of CYP2C11 in rats and CYP3A in humans (Li et al., 1993).

It may also be noted that nonpesticidal inhibitors of CYP isoforms may affect the metabolism and toxicity of pesticides. For example, Agyeman and Sultatos (1998) showed that the H2-blocker cimetidine caused a moderate increase in the toxicity of parathion but did not affect the toxicity of paraoxon, an effect brought about by the inhibition of CYP isoforms.

BIPHASIC EFFECTS: INHIBITION AND INDUCTION

Many inhibitors of mammalian CYP-dependent monooxygenase activity can also act as inducers. Generally, inhibition of microsomal monooxygenase activity is fairly rapid and involves a direct interaction with the cytochrome, whereas induction is a slower process. Therefore, after a single injection an initial decrease due to inhibition is followed by an inductive phase. As the compound and its metabolites are eliminated, the levels of activity return to control values.

Some of the best examples of compounds showing such biphasic effects are MDP compounds, including the pesticide synergists piperonyl butoxide and sesamex, and the secondary plant compounds safrole and isosafrole. The effect of MDP compounds on CYP activity is an initial inhibition of activity followed by an increase above control levels (Kamienski and Murphy, 1971; Kinsler et al., 1990; Philpot and Hodgson, 1971–1972). The inhibitory effect of MDP compounds has been attributed to the formation of a stable inhibitory metabolite complex between the heme iron of the CYP and the carbene species formed when water is cleaved from the hydroxylated methylene carbon of the MDP compound (Dahl and Hodgson, 1979). Because CYP combined with MDP compounds in an inhibitory complex cannot interact with CO, the CYP titer, as determined by the method of Omura and Sato (1964) (dependent upon CO binding to reduced cytochrome), reflects the biphasic effect.

MDP exposure induces several hepatic CYP isoforms in mice, including Cyp1a1, Cyp1a2, and Cyp2b10 (Adams et al., 1993a,b; Cook and Hodgson, 1985, 1986; Lewandowski et al., 1990; Ryu et al., 1995, 1996, 1997; Ryu and Hodgson, 1999). A number of studies have been published regarding the effects of MDP compounds on mammalian liver enzymes (for reviews see Adams et al., 1995; Hodgson et al., 1995a,b).

It is apparent from extensive reviews of the induction of monooxygenase activity by xenobiotics that many compounds other than methylenedioxyphenyl compounds have the same biphasic effect. It may be that any synergist that functions by inhibiting microsomal monooxygenase activity could also induce this activity on longer exposure, resulting in a biphasic curve, as described previously for methylene-dioxyphenyl compounds. This curve has been demonstrated for NIA 16824 (2-methylpropyl-2-propynyl phenylphosphonate) and WL 19255 (5,6-dichloro-1,2,3-benzothiadiazole), although the results were less marked with R05-8019 [2-(2,4,5-trichlorophenyl)propynyl ether] and MGK 264 [N-(2-ethylhexyl)-5-norbornene-2,3-dicarboximide].

ACTIVATION

Activation, as distinct from induction, is a stimulatory effect on enzyme activity caused by an interaction at the active site of the enzyme and/or an allosteric effect on enzyme protein conformation. As a consequence, activation tends to be rapid. Induction (discussed earlier), on the other hand, involves the synthesis of new enzyme and tends to be slower than activation.

Although activation of CYP enzyme activity is less frequently encountered, and less well understood, than either inhibition or induction, it has been known for some time. Enhancement, by acetone, of the hepatic microsomal p-hydroxylation was first reported in 1968 (Anders, 1968). Flavone and benzoflavone both stimulate benzo(a) pyrene metabolism by rabbit liver CYPs, the extent of stimulation depending on the CYP isoform involved (Huang et al., 1981). 6β-Hydroxylation of testosterone by the human isoform CYP3A4 is significantly increased by incubation of the enzyme with pyridostigmine bromide (Usmani et al., 2003), and Buratti and Testai (2007) have presented evidence for the autoactivation of CYP3A4 during dimethoate metabolism. More recently, Cho et al. (2007) have shown that chlorpyrifos oxon significantly activates the production of 1-naphthol, 2-naphthol, *trans*-1,2-dihydronaphthalenediol, and 1,4-naphthoquinine from naphthalene by human liver microsomes. Further, it was shown that production of naphthalene metabolites by CYPs 2C8, 2C9, 2C19, 2D6, 3A4, 3A5, and 3A7 was activated by chlorpyrifos oxon, while the production of naphthalene metabolites by CYPs 1A1, 1A2, 1B1, and 2B6 was inhibited by chlorpyrifos oxon.

Activation effects on CYP metabolism of the insect repellent DEET (*N*,*N*-diethyl-*m*-toluamide) were also noted (Cho et al., 2007). Chlorpyrifos oxon inhibited the formation of *N*,*N*-diethyl-*m*-hydroxymethylbenzamide from DEET by human liver microsomes while stimulating the formation from DEET of *N*-ethyl-*m*-toluamide. This was reflected by the finding that CYP2B6, the principal isoform for *N*,*N*-diethyl-*m*-hydroxymethylbenzamide production, was inhibited by chlorpyrifos oxon, while CYP3A4, the principal isoform for *N*-ethyl-*m*-toluamide production, was activated.

HEPATOTOXICITY

Hepatotoxicity has frequently been observed as a consequence of xenobiotic exposure. Although XMEs and metabolic interactions may not be directly involved in hepatotoxicity, some consideration should be given to this phenomenon, since loss of liver function may well give rise to consequences and interactions not seen in the intact liver. There have been a number of studies of enzyme induction in isolated hepatocytes from both surrogate animals and humans (see Induction of Microsomal Enzyme Activity), but studies involving the toxic effects of pesticides on the hepatocytes of surrogate animals have been relatively rare. The availability of the liver-derived HepG2 cell line and the more recent availability of human hepatocytes have, however, made new approaches to this phenomenon possible.

Typically, cell viability and cytotoxicity are measured by the use of the trypan blue exclusion method and the release of adenylate kinase into the medium. Apoptosis, or programmed cell death, is examined by measuring the induction of caspase-3/7 (see Das et al., 2006, for a description of these methods). In both HepG2 cells and human hepatocytes the following pesticides caused cell death, release of adenylate kinase, and induction of caspase: fipronil and fipronil sulfone (Das et al., 2006), deltamethrin and permethrin (Das et al., 2008a), DEET (Das et al., 2008b), and chlorpyrifos (Das et al., 2008b). The most potent of these is fipronil, while the least potent is DEET. Since fipronil sulfone is more active than fipronil, the CYP-dependent monooxygenation of fipronil may be considered an activation reaction.

CONCLUSIONS

Knowledge of the metabolism of pesticides is essential and further knowledge is still needed for several reasons, including the development of more selective insecticides and for providing, in part, the fundamental basis for science-based risk assessments for human and environmental health. Since multiple exposures tend to be the rule rather than the exception, knowledge of metabolic interactions is a vital adjunct to the risk analysis process, one that is still inadequately understood or considered. Until relatively recently, and as a matter of necessity, this research was carried out almost exclusively on experimental animals, and the results, particularly in the case of human health risk assessments, were extrapolated to humans. Although much essential background will continue to be obtained from experimental animals, because of the new techniques of molecular biology and the availability of human cells, human cell fractions, and recombinant human enzymes, it is now possible to work directly on human biotransformations and metabolic interactions. Molecular techniques also permit the study of genetic polymorphisms that will enable us to identify populations at increased risk and enable studies to be carried out at the level of specific isoforms of the XMEs involved. The interaction of pesticides and clinical drugs, although long a subject for speculation, has been the subject of little investigation. The work of Di Consiglio et al. (2005) on the interaction between imipramine and organophosphorothionates makes it clear that much more work is needed on this problem.

Thus the study of pesticide metabolism and metabolic interactions of pesticides has entered a new molecular era that will be fascinating as well as useful.

REFERENCES

Abernathy, C. O., Hodgson, E., & Guthrie, F. E. (1971a). Structure–activity relationships on the induction of hepatic microsomal enzymes in the mouse by 1,1,1-trichloro-2,2-bis(p-chlorophenyl)ethane (DDT) analogs. *Biochem. Pharmacol.*, *20*, 2385–2393.

Abernathy, C. O., Philpot, R. M., Guthrie, F. E., & Hodgson, E. (1971b). Inductive effects of 1,1,1-trichloro-2,2-bis(p-chlorophenyl)ethane (DDT), phenobarbital and benzpyrene on microsomal cytochrome P-450, ethyl isocyanide spectra and metabolism in vivo of zoxazolamine and hexobarbital in the mouse. *Biochem. Pharmacol.*, *20*, 2395–2400.

Adams, N. H., Levi, P. E., & Hodgson, E. (1993a). Differences in induction of three P450 isozymes by piperonyl butoxide, sesamex, safrole, and isosafrole. *Pestic. Biochem. Physiol.*, *46*, 15–26.

Adams, N. H., Levi, P. E., & Hodgson, E. (1993b). Regulation of cytochrome P450 isozymes by methylenedioxyphenyl compounds. *Chem. Biol. Interact.*, *86*, 225–274.

Adams, N. H., Levi, P. E., & Hodgson, E. (1995). Regulation of cytochrome P450 isozymes by methylenedioxyphenyl compounds—an updated review of the literature. *Rev. Biochem. Toxicol.*, *11*, 205–222.

Aggeler, P. M., & O'Reilly, R. A. (1969). Effect of heptabarbital on the response to bishydroxycoumarin in man. *J. Lab. Clin. Med.*, *74*, 229–238.

Agyeman, A. A., & Sultatos, L. G. (1998). The actions of the H2-blocker cimetidine on the toxicity of the phosphorothionate insecticide parathion. *Toxicology*, *128*, 207–218.

Allen, J. W., Wolf, D. C., George, M. H., Hester, S. D., Sun, G. B., Thai, S. F., et al. (2006). Toxicity profiles in mice treated with hepatotumorigenic and non-hepatotumorigenic triazole conazole fungicides: Propiconazole, triadimefon and myclobutanil. *Toxicol. Pathol.*, *34*, 853–862.

Anders, M. W. (1968). Acetone enhancement of microsomal aniline para-hydroxylase activity. *Arch. Biochem. Biophys.*, *126*, 269–275.

Anderson, K. E., & Kappas, A. (1991). Dietary regulation of cytochrome P450. *Annu. Rev. Nutr.*, *11*, 141–167.

Baker, R. C., Coons, L. B., Mailman, R. B., & Hodgson, E. (1972). Induction of hepatic mixed function oxidases by the insecticide, Mirex. *Environ. Res.*, *5*, 418–424.

Barton, H. A., Tang, J., Sey, S. M., Stanko, J. P., Murrell, R. N., Rockett, J. C., et al. (2006). Metabolism of myclobutanil and triadimefon by human and rat cytochrome P450 enzymes and liver microsomes. *Xenobiotica*, *36*, 793–806.

Batt, A. M., Siest, G., Magdalou, J., & Galteau, M.-M. (1992). Enzyme induction by drugs and toxins. *Clin. Chim. Acta*, *209*, 109–121.

Berger, C. W., & Sultatos, L. G. (1996). Mixed inhibition of mouse hepatic microsomal 2-hydroxylation of 17β-estradiol by fenitrothion. *Toxicologist*, *30*, 52–53 (Abstract 272).

Birnbaum, L. (1994). Endocrine effects of prenatal exposure to PCBs, dioxins, and other xenobiotics: Implications for policy and future research. *Environ. Health Perspect.*, *102*, 676–679.

Brown, R. R., Miller, J. A., & Miller, E. C. (1954). The metabolism of methylated aminoazo dyes. IV. Dietary factors enhancing demethylation *in vitro*. *J. Biol. Chem.*, *209*, 211–222.

Buratti, F. M., & Testai, E. (2007). Evidences for CYP3A4 autoactivation in the desulfuration of dimethoate by the human liver. *Toxicology*, *241*, 33–46.

Butler, A. M., & Murray, M. (1993). Inhibition and inactivation of constitutive cytochromes P450 in rat liver by parathion. *Mol. Pharmacol.*, *43*, 902–908.

Butler, A. M., & Murray, M. (1997). Biotransformation of parathion in human liver: Participation of CYP3A4 and its inactivation during microsomal parathion oxidation. *J. Pharmacol. Exp. Ther.*, *280*, 966–973.

Butler, L. E., & Dauterman, W. C. (1988). The effect of dietary protein levels on xenobiotic biotransformation in F344 male rats. *Toxicol. Appl. Pharmacol.*, *95*, 301–310.

Butler, L. E., & Dauterman, W. C. (1989). Sensitivity of selected drug bio-transformation enzymes to dietary protein levels in adult F344 male rats. *J. Biochem. Toxicol.*, *4*, 71–72.

Casabar, R. C. T., Das, P. C., DeKrey, G. K., Gardiner, C. S., Cao, Y., Rose, R. L., et al. (2010). Endosulfan induces CYP2B6 and CYP3A4 by activating the pregnane X receptor. *Toxicol. Appl. Pharmacol.*, *245*, 335–343.

Chadwick, R. W., & Freal, J. J. (1972). Comparative acceleration of lindane metabolism to chlorophenols by pretreatment of rats with lindane or with DDT and lindane. *Food Cosmet. Toxicol.*, *10*, 789–795.

Chadwick, R. W., Cranmer, M. F., & Peoples, A. J. (1971). Comparative stimulation of HCH metabolism by pretreatment of rats with γHCH, DDT, and DDT + γHCH. *Toxicol. Appl. Pharmacol.*, *18*, 685–695.

Choi, J., Hodgson, E., & Rose, R. L. (2004). Inhibition of trans-permethrin hydrolysis in human liver fractions by chlorpyrifos oxon and carbaryl. *Drug Metab. Drug Interact.*, *20*, 233–246.

Cohen, S. D. (1984). Mechanisms of toxicological interactions involving organophosphate insecticides. *Fundam. Appl. Toxicol.*, *4*, 315–324.

Colborn, T., vom Saal, F. S., & Soto, A. M. (1993). Developmental effects of endocrine-disrupting chemicals in wildlife and humans. *Environ. Health Perspect.*, *10*, 93–103.

Coldwell, B. B., & Zawidzka, L. (1968). Effect of acute administration of acetylsalicylic acid on the prothrombin activity of bishydroxy-coumarin-treated rats. *Blood*, *32*, 945–949.

Conney, A. H. (1967). Pharmacological implications of microsomal enzyme induction. *Pharmacol. Rev.*, *19*, 317–366.

Conney, A. H., Miller, E. C., & Miller, J. A. (1957). Substrate-induced synthesis and other properties of benzpyrene hydroxylase in rat liver. *J. Biol. Chem.*, *228*, 753–766.

Conney, A. H., Welch, R. M., Kuntzman, R., & Burns, J. J. (1967). Effects of pesticides on drug and steroid metabolism. *Clin. Pharmacol. Ther.*, *8*, 2–10.

Conney, A. H., Buening, M. K., Pantuck, C. B., Fortner, J. G., Anderson, K. E., & Kappas, A. (1979). Regulation of human drug metabolism by dietary factors. *Environ. Chem. Enzyme Funct. Hum. Dis. Ciba Found. Symp.*, *76*, 147–167.

Cook, J. C., & Hodgson, E. (1985). The induction of cytochrome P-450 by isosafrole and related methylenedioxyphenyl compounds. *Chem. Biol. Interact.*, *54*, 299–355.

Cook, J. C., & Hodgson, E. (1986). The induction of cytochrome P-450 in congenic C57/BL/6J mice by isosafrole: Lack of correlation with the Ah locus. *Chem. Biol. Interact.*, *58*, 223–240.

Cranmer, M. F. (1970). Effect of diphenylhydantoin on storage in the rat. *Toxicol. Appl. Pharmacol.*, *17*, 315.

Cranmer, M. F., Peoples, A., & Chadwick, R. (1972). Biochemical effects of repeated administration of *p,p*-DDT on the squirrel monkey. *Toxicol. Appl. Pharmacol.*, *21*, 98–101.

Crouch, L. S., & Ebel, R. E. (1987). Benzo(a)pyrene metabolism in the Mongolian gerbil: Influence of chlordecone and mirex induction. *Xenobiotica*, *17*, 859–867.

Cullen, S. I., & Catalano, P. M. (1967). Griseofulvin–warfarin antagonism. *J. Am. Med. Assoc.*, *199*, 582–583.

Dahl, A. R., & Hodgson, E. (1979). The interaction of aliphatic analogs of methylene-dioxyphenyl compounds with cytochromes P450 and P430. *Chem. Biol. Interact.*, *27*, 163–175.

Dai, D., Rose, R. L., & Hodgson, E. (1998). Toxicology of environmentally persistent chemicals: Mirex and chlordecone. *Rev. Toxicol.*, *2*, 477–499.

Dalton, S. R., Miller, R. T., & Meyer, S. A. (2003). The herbicide metolachlor induces liver cytochrome P450s 2B1/2 and 3A1/2, but not thyroxine–uridine dinucleotide phosphate glucuronosyltransferase and associated thyroid gland activity. *Int. J. Toxicol.*, *22*, 287–295.

Das, P. C., Cao, Y., Cherrington, N., Hodgson, E., & Rose, R. L. (2006). Fipronil induces CYP isoforms and cytotoxicity in human hepatocytes. *Chem. Biol. Interact.*, *164*, 200–214.

Das, P. C., Streit, T. M., Cao, Y., Rose, R. L., Cherrington, N., Ross, M. K., et al. (2008a). Pyrethroids: Cytotoxicity and induction of CYP isoforms in human hepatocytes. *Drug Metab. Drug Interact.*, *23*, 211–236.

Das, P. C., Cao, Y., Rose, R. L., Cherrington, N., & Hodgson, E. (2008b). Enzyme induction and cytotoxicity in human hepatocytes by chlorpyrifos and N,N-diethyl-m-toluamide (DEET). *Drug Metab. Drug Interact.*, *23*, 237–260.

Dehn, P. F., Allen-Mocherie, S., Karek, J., & Thenappen, A. (2005). Organochlorine insecticides impacts on human HepG2 cytochrome P4501A, 2B activities and glutathione levels. *Toxicol. in Vitro*, *19*, 261–273.

Deitrich, R. A. (1971). Genetic aspects of increase in rat liver aldehyde dehydrogenase induced by phenobarbital. *Science*, *173*, 334–336.

Denison, M. S., & Whitlock, J. P. Jr., (1995). Xenobiotic-inducible transcription of cytochrome P450 genes. *J. Biol. Chem.*, *270*, 18175–18178.

Di Consiglio, E., Meneguz, A., & Testai, E. (2005). Organophosphorothionate pesticides inhibit the bioactivation of imipramine by human hepatic cytochrome P450s. *Toxicol. Appl. Pharmacol.*, *205*, 237–246.

Donaldson, W. E. (1994). Nutritional factors. In E. Hodgson & P. E. Levi (Eds.), *Introduction to biochemical toxicology* (2nd ed., pp. 297–317) Norwalk, CT: Appleton & Lange.

DuBois, K. P. (1969). Combined effects of pesticides. *Can. Med. Assoc. J.*, *100*, 173–179.

DuBois, M., Plaisance, H., Thome, J. P., & Kremers, P. (1996). Hierarchical cluster analysis of environmental pollutants through P450 induction in cultured hepatic cells. *Ecotoxicol. Environ. Saf.*, *34*, 205–215.

Fabacher, D. L., & Hodgson, E. (1976). Induction of hepatic mixed function oxidase enzymes in adult and neonatal mice by kepone and mirex. *Toxicol. Appl. Pharmacol.*, *38*, 71–77.

Fabacher, D. L., Kulkarni, A. P., & Hodgson, E. (1980). Pesticides as inducers of hepatic drug-metabolizing enzymes. I. Mixed function oxidase activity. *Gen. Pharmacol.*, *11*, 429–435.

Fine, B. C., & Molloy, J. O. (1964). Effects of insecticide synergists on duration of sleep induced in mice by barbiturates. *Nature*, *204*, 789–790.

Fouse, B. L., & Hodgson, E. (1987). Effects of chlordecone and mirex on the acute hepatotoxicity of acetaminophen in mice. *Gen. Pharmacol.*, *18*, 623–630.

Fouts, J. R. (1963). Factors influencing the metabolism of drugs in liver microsomes. *Ann. N.Y. Acad. Sci.*, *104*, 875–880.

Gaughan, L. C., Engel, J. L., & Casida, J. E. (1980). Pesticide interactions: Effects of organophosphorus insecticides on the metabolism, toxicity, and persistence of selected pyrethroid insecticides. *Pestic. Biochem. Physiol.*, *14*, 81–85.

Gelboin, H. V., & Conney, A. H. (1968). Antagonism and potentiation of drug action. In E. Boyland & R. Goulding (Eds.), *Modern trends in toxicology* (pp. 175–195). London: Butterworths.

Gielen, J. E., & Nebert, D. W. (1971). Microsomal hydroxylase induction in liver cell culture by phenobarbital, polycyclic hydrocarbons, and *p,p*-DDT. *Science*, *172*, 167–169.

Gillette, J. R., Davis, D. C., & Sasame, H. A. (1972). Cytochrome P450 and its role in drug metabolism. *Annu. Rev. Pharmacol.*, *12*, 57–84.

Goetz, A. K., Bao, W., Ren, H., Schmid, J. E., Tully, D. B., Wood, C., et al. (2006). Gene expression profiling in the liver of CD-1 mice to characterize the hepatotoxicity of triazole fungicides. *Toxicol. Appl. Pharmacol.*, *215*, 274–284.

Goldstein, J. A., Hickan, P., & Kimbrough, R. D. (1973). Effects of purified and technical piperonyl butoxide on drug-metabolizing enzymes and ultrastructure of rat liver. *Toxicol. Appl. Pharmacol.*, *26*, 444–453.

Granick, S. (1965). Hepatic porphyria and drug-induced or chemical porphyria. *Ann. N.Y. Acad. Sci.*, *123*, 188–197.

Guillette, L. J., Jr., Gross, T. S., Masson, G. R., Matter, J. M., Percival, H. F., & Woodward, A. R. (1994). Developmental abnormalities of the gonad and abnormal sex hormone concentrations in juvenile alligators from contaminated and control lakes in Florida. *Environ. Health Perspect.*, *102*, 680–688.

Guzelian, P. S., Vramian, G., Boylan, J. J., et al. (1980). Liver structure and function in patients poisoned with chlordecone (kepone). *Gastroenterology*, *78*, 206–213.

Halpert, J., Hammond, D., & Neal, R. M. (1980). Inactivation of purified rat liver cytochrome P-450 during the metabolism of parathion (diethyl *p*-nitrophenyl phosphorothionate). *J. Biol. Chem.*, *255*, 1080–1089.

Hanioka, N., Watanabe, K., Yoda, R., & Ando, M. (2002). Effect of alachlor on hepatic cytochrome P450 enzymes in rats. *Drug Chem. Toxicol.*, *25*, 25–37.

Hart, L. G., & Fouts, J. R. (1963). Effects of acute and chronic DDT administration on hepatic microsomal drug metabolism in the rat. *Proc. Soc. Exp. Biol. Med.*, *114*, 388–392.

Heder, A. F., Hirsch-Ernst, K. I., Bauer, D., Kahl, G. F., & Desel, H. (2001). Induction of cytochrome P-450 2B1 by pyrethroids in primary rat hepatocyte cultures. *Biochem. Pharmacol.*, *62*, 71–79.

Hinderling, P. H., & Garrett, E. R. (1977). Pharmacokinetics of β-methyl-digoxin in healthy humans. III. Pharmacodynamic correlations. *J. Pharmacol. Sci.*, *66*, 326–329.

Hodgson, E. (1974). Comparative studies of cytochrome P450 and its interaction with pesticides. In M. A. Q. Khan & J. P. Bederka, Jr. (Eds.), *Survival in toxic environments* (pp. 213–260). New York: Academic Press.

Hodgson, E. (1999). Induction and inhibition of pesticide-metabolizing enzymes: Roles in synergism of pesticides and pesticide action. *Toxicol. Ind. Health*, *15*, 6–11.

Hodgson, E., & Kulkarni, A. P. (1974). Interactions of pesticides with cytochrome P450. *ACS Symp. Ser.*, *2*, 14–38.

Hodgson, E., & Levi, P. E. (1996). Pesticides: An important but underused model of the environmental health sciences. *Environ. Health Perspect.*, *104*, 97–106.

Hodgson, E., & Levi, P. E. (1998). Interactions of piperonyl butoxide with cytochrome P450. In D. Glynne Jones (Ed.), *Piperonyl butoxide, the insecticide synergist*. San Diego: Academic Press.

Hodgson, E., & Meyer, S. A. (1997). Pesticides. In I. G. Sipes, C. A. McQueen, & A. J. Gandolfi (Series Eds.) & R. S. McCluskey & D. L. Earnest (Vol. Eds.), *Comprehensive toxicology. Hepatic and gastrointestinal toxicology*. Oxford, UK: Pergamon.

Hodgson, E., and Meyer, S. A. (2010). Pesticides and hepatotoxicity. In R. Roth & P. E. Ganey (Series Eds.), & C. A. McQueen (Ed.), *Comprehensive toxicology*. Vol. 9. *Hepatic toxicology*. New York: Elsevier.

Hodgson, E., & Philpot, R. M. (1974). Interaction of methylenedioxyphenyl (1,3-benzodioxole) compounds with enzymes and their effects on mammals. *Drug. Metab. Rev.*, *3*, 231–301.

Hodgson, E., Kulkarni, A. P., Fabacher, D. L., & Robacker, K. M. (1980). Induction of hepatic drug metabolizing enzymes in mammals by pesticides: A review. *J. Environ. Sci. Health B*, *15*, 723–754.

Hodgson, E., Rose, R. L., Ryu, D.-Y., Falls, G., Blake, B. L., & Levi, P. E. (1995). Pesticide-metabolizing enzymes. *Toxicol. Lett.*, *82–83*, 73–81.

Hodgson, E., Ryu, D-Y., Adams, N., & Levi, P. E. (1995). Biphasic responses in synergistic reactions. *Toxicology*, *105*, 211–216.

Hodgson, E., Cherrington, N., Coleman, S. C., Liu, S., Falls, J. G., Cao, Y., et al. (1998). Flavin-containing monooxygenase and cytochrome P450 mediated metabolism of pesticides: From mouse to human. *Rev. Toxicol.*, *2*, 231–243.

Hoffman, D. G., Worth, H. M., Emmerson, J. L., & Anderson, R. C. (1968). Stimulation of hepatic microsomal drug-metabolizing enzymes by α-α-bis(*p*-chlorophenyl)-3-pyridinemethanol and a method for determining no-effect levels in rats. *Toxicol. Appl. Pharmacol.*, *12*, 464–472.

Huang, M., Johnson, E. F., Muller-Eberhard, U., Koop, D. R., & Conney, A. H. (1981). Specificity in the activation and inhibition by flavinoids of benzo(a)pyrene hydroxylation by cytochrome P-450 isozymes from rabbit liver microsomes. *J. Biol. Chem.*, *256*, 10897–10901.

Jager, E., Platzed, T., & Helge, H. (1980). The ^{13}C aminopyrine breath test for the assessment of liver function in children. In E. Gladtke & G. Heimann (Eds.), *Pharmacokinetics* (pp. 271–276). Stuttgart: Fischer.

Jensen, I. M., & Whatling, P. (2010). Malathion: A review of toxicology. In *Haye's handbook of pesticide toxicology*. San Diego: Elsevier (Chap. 71).

Johri, A., Yadav, S., Singh, R. L., Dhawan, A., Ali, M., & Parmar, D. (2006). Long lasting effects of prenatal exposure to deltamethrin on cerebral and hepatic cytochrome P450s and behavioral activity in rat offspring. *Eur. J. Pharmacol.*, *544*, 58–68.

Johri, A., Dhawan, A., Singh, R. L., & Parmar, D. (2008). Persistence in alterations in the ontogeny of cerebral and hepatic cytochrome P450s following prenatal exposure to low doses of lindane. *Toxicol. Sci.*, *101*, 331–340.

Joo, H., Choi, K., Rose, R. L., & Hodgson, E. (2007). Inhibition of fipronil and n-nonane metabolism in human liver microsomes and human cytochrome P450 (CYP) isoforms by chlorpyrifos. *J. Biochem. Metab. Toxicol.*, *21*, 76–80.

Kadlubar, F. F., Butler, M. A., Kaderlik, K. R., Chou, H. C., & Lang, N. P. (1992). Polymorphisms for aromatic amine metabolism in humans: Relevance for human carcinogenesis. *Environ. Health Perspect.*, *98*, 69–74.

Kamataki, T., & Neal, R. A. (1976). Metabolism of diethyl-*p*-nitrophenyl phosphorothionate (parathion) by a reconstituted mixed function oxidase enzyme system: Studies of its covalent binding of the sulfur atom. *Mol. Pharmacol.*, *12*, 933–944.

Kamienski, F. X., & Murphy, S. D. (1971). Biphasic effects of methylene-dioxyphenyl synergists on the action of hexobarbital and organophosphate insecticides in mice. *Toxicol. Appl. Pharmacol.*, *18*, 883–894.

Kaminisky, L. S., Piper, L. J., McMartin, D. N., & Fasco, M. J. (1978). Induction of hepatic microsomal cytochrome P450 by mirex and kepone. *Toxicol. Appl. Pharmacol.*, *43*, 327–338.

Khan, M. A. Q. (1984). Induction of drug metabolizing enzymes. In F. Matsumura (Ed.), *International encyclopedia of pharmacology and therapeutics*. Section 113, *Differential toxicities of insecticides and halogenated aromatics* (p. 113). New York: Pergamon.

Kinoshita, F. K., & DuBois, K. P. (1967). Effects of substituted urea herbicides on activity of hepatic microsomal enzymes. *Toxicol. Appl. Pharmacol.*, *10*, 410.

Kinoshita, F. K., Frawley, J. P., & DuBois, K. P. (1966). Quantitative measurement of induction of hepatic microsomal enzymes by various dietary levels of DDT and toxaphene in rats. *Toxicol. Appl. Pharmacol.*, *9*, 505–513.

Kinsler, S., Levi, P. E., & Hodgson, E. (1990). Relative contributions of cytochrome P450 and flavin-containing monooxygenases to the microsomal oxidation of phorate following treatment of mice with phenobarbital, hydrocortisone, acetone, and piperonyl butoxide. *Pestic. Biochem. Physiol.*, *37*, 174–181.

Kiyosawa, N., Kwekel, J. C., Burgoon, L. D., Williams, K. J., Tashiro, C., Chittim, B., et al. (2008). o,p-DDT elicits PXR/CAR-, not ER-mediated responses in the immature ovariectomized rat liver. *Toxicol. Sci.*, *101*, 350–363.

Kobliakov, V., Popova, N., & Rossi, L. (1991). Regulation of the expression of sex-specific isoforms of cytochrome P450 in rat liver. *Eur. J. Biochem.*, *195*, 588–591.

Kolmodin, B., Azarnoff, D. L., & Sjoquist, F. (1969). Effect of environmental factors on drug metabolism: Decreased plasma half-life of antipyrine in workers exposed to chlorinated hydrocarbon insecticides. *Clin. Pharmacol. Ther.*, *10*, 638–642.

Kolmodin-Hedman, B. (1973). Decreased plasma half-life of phenylbutazone in workers exposed to chlorinated pesticides. *Eur. J. Clin. Pharmacol.*, *5*, 195–198.

Koransky, W., Portig, J., Vohland, H. W., & Klempau, I. (1964). Activation of microsomal enzymes by isomers of hexachlorocyclohexane: Its influence on scilliroside poisoning in the rat. *Naunyn-Schmiedebergs Arch. Exp. Pathol. Pharmakol.*, *247*, 61–78.

Kreiss, K., Zack, M. M., Kimbrough, R. D., Needham, L. L., Smrek, A. L., & Jones, B. T. (1981). Cross-sectional study of a community with exceptional exposure to DDT. *J. Am. Med. Soc.*, *2245*, 1926–1930.

Krishnaswamy, K., Ushasri, V., & Nadamuni Naidu, A. (1981). The effect of malnutrition on the pharmacokinetics of phenylbutazone. *Clin. Pharmacol.*, *6*, 152–159.

Kulkarni, A. P., & Hodgson, E. (1984a). Metabolism of insecticides by the microsomal mixed function oxidase systems. In F. Matsumura (Ed.), *International encyclopedia of pharmacology and therapeutics.* Section 113, *Differential toxicities of insecticides and halogenated aromatics* (pp. 27–128). New York: Pergamon.

Kulkarni, A. P., & Hodgson, E. (1984b). The metabolism of insecticides: The role of monooxygenase enzymes. *Annu. Rev. Pharmacol. Toxicol., 24,* 19–42.

Kulkarni, A. P., Fabacher, D. L., & Hodgson, E. (1980). Pesticides as inducers of hepatic drug metabolizing enzymes. II. Glutathione S-transferases. *Gen. Pharmacol., 11,* 437–441.

Lapadula, D. M., Carrington, C. D., & Abou-Donia, M. B. (1984). Induction of hepatic microsomal cytochrome P450 and inhibition of brain, liver and plasma esterases by an acute dose of S,S,S-tri-n-butyl phosphorotrithioate (DEF) in the adult hen. *Toxicol. Appl. Pharmacol., 73,* 300–310.

Leibman, K. C. (1968). Actions of insecticides on drug activity. *Int. Anesthesiol. Clin., 6,* 251–260.

Lemaire, G., Mnif, W., Pascussi, J. M., Pillon, A., Rabenoelina, F., Fenet, H., et al. (2006). Identification of new human pregnane X receptor ligands among pesticides using a stable reporter cell system. *Toxicol. Sci., 91,* 501–509.

Levi, P. E., Hollingworth, R. M., & Hodgson, E. (1988). Differences in oxidative dearylation and desulfuration of fenitrothion by cytochrome P-450 isozymes and in the subsequent inhibition of monooxygenase activity. *Pestic. Biochem. Physiol., 32,* 224–231.

Levi, P. E., Rose, R. L., Adams, N. H., & Hodgson, E. (1992). Induction of cytochrome P450 4A1 in mouse liver by the herbicide synergist tridiphane. *Pestic. Biochem. Physiol., 44,* 9–14.

Lewandowski, M., Levi, P. E., & Hodgson, E. (1989). Induction of cytochrome P-450 isozymes by mirex and chlordecone. *J. Biochem. Toxicol., 4,* 195–199.

Lewandowski, M., Chui, Y. C., Levi, P. E., & Hodgson, E. (1990). Differences in induction of hepatic cytochrome P450 isozymes in mice by eight methylenedioxyphenyl compounds. *J. Biochem. Toxicol., 5,* 47–55.

Li, H. C., Mani, C., & Kupfer, D. (1993). Reversible and time-dependent inhibition of the hepatic cytochrome P450 steroidal hydroxylases by the proestrogenic pesticide methoxychlor in rat and human. *J. Biochem. Toxicol., 8,* 195–206.

Li, H. C., Dehal, S. S., & Kupfer, D. (1995). Induction of the hepatic CYP2B and CYP3A enzymes by the proestrogenic pesticide methoxychlor and by DDT in the rat. Effects on methoxychlor metabolism. *J. Biochem. Toxicol., 10,* 51–61.

Martin, M. T., Brennan, R. J., Hu, W. Y., Ayanoglu, E., Lau, C., Ren, H. Z., et al. (2007). Toxicogenomic study of triazole fungicides and perfluoroalkyl acids in rat livers predicts toxicity and categorizes chemicals based on mechanisms of toxicity. *Toxicol. Sci., 97,* 595–613.

Matsubara, T., Noracharttiyapot, W., Toriyabe, T., Yoshinari, K., Nagata, K., & Yamazoe, Y. (2007). Assessment of human pregnane X receptor involvement in pesticide-mediated activation of CYP3A4 gene. *Drug Metab. Dispos., 35,* 728–733.

Matthews, M. S., & Devi, K. S. (1994). Effect of chronic exposure of pregnant rats to malathion and/or estrogen and/or progesterone on xenobiotic metabolizing enzymes. *Pestic. Biochem. Physiol., 48,* 110–122.

Medina-Diaz, I. M., & Elizondo, G. (2005). Transcriptional induction of CYP3A4 by o,p-DDT in HepG2 cells. *Toxicol. Lett., 157,* 41–47.

Medina-Diaz, I. M., Arteaga, G., de Leon, M. B., Cisneros, B., Sierra-Santoyo, A., Vega, L., et al. (2007). Pregnane X receptor-dependent induction of the CYP3A4 gene by o,p'-1,1,1-trichloro-2,2-bis(p-chlorophenyl) ethane. *Drug Metab. Dispos., 35,* 95–102.

Mehta, S., Kalshi, H. K., Jayaraman, S., & Mathur, V. S. (1975). Chloramphenicol metabolism in children with protein-calorie malnutrition. *Am. J. Clin. Nutr., 28,* 977–981.

Mehta, S., Nain, C. K., Sharma, B., & Mathur, V. S. (1982). Disposition of four drugs in malnourished children. *Drug-Nutr. Int., 1,* 205–211.

Miller, E. C., Miller, J. A., & Conney, A. H. (1954). On the mechanism of the methylcholanthrene inhibition of carcinogenesis by 39-methyl-4-dimethylaminoazobenzene. *Cancer Res., 51,* 32.

Moody, D. E., & Hammock, B. D. (1987). The effect of tridiphane (2-(3,5-dichlorophenyl)-2-(2,2,2-trichloroethyl)oxirane) on hepatic epoxide-metabolizing enzymes: Indications of peroxisome proliferation. *Toxicol. Appl. Pharmacol., 89,* 37–46.

Moody, D. E., Gibson, G. G., Grant, D. F., Magdalou, J., & Rao, M. S. (1992). Peroxisome proliferators, a unique set of drug-metabolizing enzyme inducers. *Drug Metab. Dispos.*, *20*, 779–791.

Moreland, D. E., Novitsky, W. P., & Levi, P. E. (1989). Selective inhibition of cytochrome P450 isozymes by the herbicide synergist tridiphane. *Pestic. Biochem. Physiol.*, *35*, 42–49.

Morelli, M. A., & Nakatsugawa, T. (1978). Inactivation *in vitro* of microsomal oxidases during parathion metabolism. *Biochem. Pharmacol.*, *27*, 293–299.

Morisseau, C., Derbel, M., Lane, T. R., Stoutamire, D., & Hammock, B. D. (1991). Differential induction of hepatic drug-metabolizing enzymes by fenvaleric acid in male rats. *Toxicol. Sci.*, *52*, 148–153.

Murphy, S. D., & DuBois, K. P. (1957). Quantitative measurement of inhibition of the enzymatic detoxification of malathion by EPN (ethyl *p*-nitrophenyl thionobenzene phosphonate). *Proc. Soc. Exp. Biol. Med.*, *96*, 813–818.

Neal, R. A. (1985). Thiono-sulfur compounds. In M. W. Anders (Ed.), *Bioactivation of foreign compounds* (pp. 519–540). New York: Academic Press.

Neal, R. A., & Halpert, J. (1982). Toxicology of thionosulfur compounds. *Annu. Rev. Pharmacol. Toxicol.*, *22*, 321–339.

Neal, R. A., Swahate, R., Halpert, J., & Kametaki, T. (1983). Chemically reactive metabolism as suicide enzyme inhibitors. *Drug Metab. Rev.*, *14*, 49–59.

Nebert, D. W., & Jensen, N. M. (1979). The Ah locus: Genetic regulation of the metabolism of carcinogens, drugs and other environmental chemicals by cytochrome P450 mediated monooxygenases. *CRC Crit. Rev. Biochem.*, *6*, 401–437.

Okey, A. B. (1990). Enzyme induction in the cytochrome P450 system. *Pharmacol. Ther.*, *45*, 141.

Okey, A. B., Roberts, E. A., Harper, P. A., & Denison, M. S. (1986). Induction of drug-metabolizing enzymes: Mechanisms and consequences. *Clin. Biochem.*, *19*, 132–142.

Omura, T., & Sato, R. (1964). The carbon monoxide binding pigment of liver microsomes. *J. Biol. Chem.*, *239*, 2370–2378.

Oropeza-Hernandez, L. F., Lopez-Romero, R., & Albores, A. (2003). Hepatic CYP1A, 2B, 2C and 3A regulation by methoxychlor in male and female rats. *Toxicol. Lett.*, *144*, 93–103.

Pantuck, E. J., Pantuck, C. B., Garland, W. A., Min, B. H., Wattenberg, L. W., & Anderson, K. E., et al. (1979). Stimulatory effect of brussel sprouts and cabbage on human drug metabolism. *Clin. Pharmacol. Ther.*, *25*, 88–95.

Pantuck, E. J., Pantuck, C. B., Anderson, K. E., Wattenberg, L. W., Conney, A. A., & Kappas, A. (1984). Effect of brussel sprouts and cabbage on drug conjugation. *Clin. Pharmacol. Ther.*, *35*, 161–169.

Paolini, M., Barillari, S., Trespidi, S., Valgimigli, L., Pedulli, G. F., & Cantelli-Forti, G. (1999). Captan impairs CYP-catalyzed drug metabolism in the mouse. *Chem. Biol. Interact.*, *123*, 149–170.

Parmar, D., Yadov, S., Dayal, M., Johri, A., Dhawan, A., & Seth, P. K. (2003). Effect of lindane on hepatic and brain cytochrome P450s and influence of P450 modulation on lindane induced neurotoxicity. *Food Chem. Toxicol.*, *41*, 1077–1087.

Philpot, R. M., & Hodgson, E. (1971–1972). The production and modification of cytochrome P450 difference spectrum by *in vivo* administration of methylenedioxyphenyl compounds. *Chem. Biol. Interact*, *4*, 185–194.

Poland, A. P., Smith, D., Kuntzman, R., Jacobson, M., & Conney, A. H. (1970). Effect of intensive occupational exposure to DDT on phenyl butazone and cortisol metabolism in human subjects. *Clin. Pharmacol. Ther.*, *11*, 724–732.

Poland, A., Goldstein, J., Hickman, P., & Burse, V. W. (1971). A reciprocal relationship between the induction of δ-aminolevulinic acid synthetase and drug metabolism produced by *m*-dichlorobenzene. *Biochem. Pharmacol.*, *20*, 1281–1290.

Price, R. J., Giddings, A. M., Scott, M. P., Walters, D. G., Capen, C. C., Osimitz, T. G., et al. (2008). Effect of pyrethrins on cytochrome P450 isoforms in cultured rat and human hepatocytes. *Toxicology*, *243*, 84–85.

Puryear, R. L., & Paulson, G. D. (1972). Effect of carbaryl (1-naphthyl *n*-methylcarbamate) on pentobarbital-induced sleeping time and some liver microsomal enzymes in White Leghorn cockerels. *Toxicol. Appl. Pharmacol.*, *122*, 621–627.

Rappolt, R. T. (1973). Use of oral DDT in three human barbiturate intoxications: CMS arousal and/or hepatic enzyme induction by reciprocal detoxicants. *Ind. Med. Surg.*, *39*, 319.

Relling, M. V., Lin, J. S., Ayers, G. D., & Evans, W. E. (1992). Racial and gender differences in *N*-acetyltransferase, xanthine oxidase and CYP1A2 activities. *Clin. Pharmacol. Ther*, *52*, 643–658.

Remmer, H. (1958). Acceleration of the destruction of Evipan under the influence of barbiturates. *Naturwissenschaften, 45,* 189. (in German)

Robacker, R. M., Kulkarni, A. P., & Hodgson, E. (1981). Pesticide induced changes in the mouse hepatic microsomal cytochrome P450 monooxygenase system and other enzymes. *J. Environ. Sci. Health B, 16,* 529–545.

Robinson, D. S., & MacDonald, M. G. (1966). The effect of phenobarbital administration on the control of coagulation achieved during warfarin therapy in man. *J. Pharmacol. Exp. Ther., 153,* 250–253.

Ronis, M. J. J., & Cunny, H. C. (1994). Physiological (endogenous) factors affecting the metabolism of xenobiotics. In E. Hodgson & P. E. Levi (Eds.), *Introduction to biochemical toxicology* (pp. 133–151) (2nd ed.). Norwalk, CT: Appleton & Lange.

Ronis, M. J. J., Ingelman-Sundberg, M., & Badger, T. M. (1994). Induction, suppression and inhibition of multiple hepatic cytochrome P450 isozymes in the male rat and bobwhite quail (*Colinus virginianus*) by ergosterol biosynthesis inhibiting fungicides (EBIDFs). *Biochem. Pharmacol., 48,* 1953–1965.

Ryu, D.-Y., & Hodgson, E. (1999). Constitutive expression and induction of CYP1B1 mRNA in the mouse. *J. Biochem. Mol. Toxicol., 13,* 249–251.

Ryu, D.-Y., Levi, P. E., & Hodgson, E. (1995). Regulation of cytochrome P-450 isozymes CYP1A1, CYP1A2 and CYP2B10 by three benzodioxole compounds. *Chem. Biol. Interact., 96,* 235–247.

Ryu, D.-Y., Levi, P. E., Fernando-Salguero, P., Gonzalez, F. J., & Hodgson, E. (1996). Piperonyl butoxide and acenaphthylene induce cytochrome P450 1A2 and 1B1 mRNA in aromatic hydrocarbon-responsive receptor knockout mouse liver. *Mol. Pharmacol., 50,* 443–446.

Ryu, D. -Y., Levi, P. E., & Hodgson, E. (1997). Regulation of hepatic CYP1A isozymes by piperonyl butoxide and acenaphthene in the mouse. *Chem. Biol. Interact., 105,* 53–63.

Schenkman, J. B., Thummel, K. E., & Favreau, L. V. (1989). Physiological and patho-physiological alterations in rat hepatic cytochrome P450s. *Drug Metab. Rev., 20,* 557–584.

Sher, S. P. (1971). Drug enzyme induction and drug interactions: Literature tabulation. *Toxicol. Appl. Pharmacol., 18,* 780–834.

Shi, X., Dick, R. A., Ford, K. A., & Casida, J. E. (2009). Enzymes and inhibitors in neonicotinoid insecticide metabolism. *Agric. Food Chem, 57,* 4861–4866.

Straw, J. A., Waters, I. W., & Fregly, M. J. (1965). Effect of *o,p*-DDD on hepatic metabolism of pentobarbital in rats. *Proc. Soc. Exp. Biol. Med., 118,* 391–394.

Street, J. C., Mayer, F. L., & Wagstaff, D. J. (1969). Ecological significance of pesticide interactions. *Ind. Med. Surg., 38,* 409–414.

Sun, G. B., Thai, S. F., Lambert, G. R., Wolf, D. C., Tully, D. B., Goetz, A. K., et al. (2006). Fluconazole-induced hepatic cytochrome P450 gene expression and enzymatic activities in rats and mice. *Toxicol. Lett., 164,* 44–53.

Sun, G. B., Grindstaff, R. D., Thai, S. F., Lambert, G. R., Tully, D. B., Dix, D. J., et al. (2007). Induction of cytochrome P450 enzymes in rat liver by two conazoles, myclobutanil and triadimefon. *Xenobiotica, 37,* 180–193.

Tang, J., Cao, Y., Rose, R. L., & Hodgson, E. (2002). In vitro metabolism of carbaryl by human cytochrome P450 and its inhibition by chlorpyrifos. *Chem. Biol. Interact., 141,* 229–241.

Tang, J., Usmani, K. A., Hodgson, E., & Rose, R. L. (2004). *In vitro* metabolism of fipronil by human and rat cytochrome P450 and its interactions with testosterone and diazepam. *Chem. Biol. Interact., 147,* 319–329.

Tully, D. B., Bao, W., Goetz, A. K., Blystone, C. R., Ren, H., Schmid, J. E., et al. (2006). Gene expression profiling in liver and testis of rats to characterize the toxicity of triazole fungicides. *Toxicol. Appl. Pharmacol., 215,* 260–273.

Tyler, C. R., Jobling, S., & Sumpter, J. P. (1998). Endocrine disruption in wildlife: A critical review of the evidence. *Crit. Rev. Toxicol., 28,* 319–361.

Ugazio, G., Burdino, E., Dacasto, M., Bosio, A., van 't Klooster, G., & Nebbia, C. (1993). Induction of hepatic drug metabolizing enzymes and interaction with carbon tetrachloride in rats after a single oral exposure to atrazine. *Toxicol. Lett., 69,* 279–288.

Upham, J., Acott, P. D., O'Regan, P., Sinal, C. J., Crocker, J. F. S., Geldenhuys, L., et al. (2007). The pesticide adjuvant, Toximul™ alters hepatic metabolism through effects on downstream targets of PPAR alpha. *Biochim. Biophys. Acta Mol. Basis Dis., 1772,* 1057–1064.

Usmani, K. A., Rose, R. L., Goldstein, J. A., Taylor, W. G., Brimfield, A. A., & Hodgson, E. (2002). In vitro human metabolism and interactions of repellent, N,N-diethyl-m-toluamide. *Drug Metab. Dispos.*, *30*, 289–294.

Usmani, K. A., Rose, R. L., & Hodgson, E. (2003). Inhibition and activation of the human liver and human cytochrome P450 3A4 metabolism of testosterone by deployment-related chemicals. *Drug Metab. Dispos.*, *31*, 384–391.

Usmani, K. A., Hodgson, E., & Rose, R. L. (2004). In vitro metabolism of carbofuran by human, mouse, and rat cytochrome P450 and interactions with chlorpyrifos, testosterone, and estradiol. *Chem. Biol. Interact.*, *150*, 221–232.

Usmani, K. A., Cho, T. M., Rose, R. L., & Hodgson, E. (2006). Inhibition of the human liver microsomal and human cytochrome P450 1A2 and 3A4 metabolism of estradiol by deployment-related and other chemicals. *Drug Metab. Dispos.*, *34*, 1606–1614.

Ward, W. O., Delker, D. A., Hester, S. D., Thai, S. F., Wolf, D. C., Allen, J. W., et al. (2006). Transcriptional profiles in liver from mice treated with hepatotumorigenic and nonhepatotumorigenic triazole con-azole fungicides: Propiconazole, triadimefon and myclobutanil. *Toxicol. Pathol.*, *34*, 863–878.

Wattenberg, L. W. (1971). Studies of polycyclic hydrocarbon hydroxylases of the intestine possibly related to cancer: Effect of diet on benzpyrene hydroxylase activity. *Cancer (Philadelphia)*, *28*, 99–102.

Welch, R. M., & Harrison, Y. (1966). Reduced drug toxicity following insecticide treatment. *Pharmacologist*, *8*, 217.

Welch, R. M., Harrison, Y., & Burns, J. J. (1967). Hemorrhagic crises in the dog caused by dicumarol–phenobarbital interaction. *Fed. Proc. Fed. Am. Soc. Exp. Biol.*, *26*, 568.

Welch, R. M., Levin, W., Kuntzman, R., Jacobson, M., & Conney, A. H. (1971). Effect of halogenated hydrocarbon insecticides on the metabolism and uterotropic action of estrogens in rats and mice. *Toxicol. Appl. Pharmacol.*, *19*, 234–246.

Wilkinson, C. F., & Denison, M. S. (1982). Pesticide interactions with bio-transformation systems. In J. E. Chambers & J. D. Yarbrough (Eds.), *Effects of chronic exposure to pesticides on animal systems* (pp. 1–24). New York: Raven Press.

Wright, A. S., Potter, D., Wooder, M. F., Donninger, C., & Greenland, R. D. (1972). Effects of dieldrin on mammalian hepatocytes. *Food Cosmet. Toxicol.*, *10*, 311–322.

Wyde, M. E., Bartelucci, E., Ueda, A., Zhang, H., Yan, B. F., Negishi, M., et al. (2003). The environmental pollutant 1,1-dichloro-2,2-bis (p-chlorophenyl) ethylene induced rat hepatic cytochrome P4502B and 3A expression through the constitutive androstane receptor and pregnane X receptor. *Mol. Pharmacol.*, *64*, 474–481.

Yu, T. L., & Varma, D. R. (1982). Pharmacokinetics, metabolism and disposition of salicylate in protein-deficient rats. *Drug Metab. Dispos.*, *10*, 147–152.

Zhao, B., Baston, D. S., Hammock, B., & Denison, M. S. (2006). Interaction of diuron and related substituted phenylureas with the Ah receptor pathway. *J. Biochem. Mol. Toxicol.*, *20*, 103–113.

Pesticide Excretion

Ernest Hodgson
North Carolina State University, Raleigh, NC, USA

Outline

INTRODUCTION

Although, in the time since the publication of the second and third editions of the *Handbook of Pesticide Toxicology* (Krieger, 2001, 2010), studies of the excretory mechanisms for pesticide excretion in vivo have received little attention, two aspects have continued to advance: cellular elimination and the use of urinary metabolites as biomarkers of pesticide exposure.

Except in simple life forms, the elimination of toxicants, including pesticides and their metabolites, is part of a specialized system that, in addition to elimination, maintains the balance of water, minerals, and other substances necessary for terrestrial life. Pesticides, again typical of toxicants in general, are taken up by the body in most cases because of their lipophilicity. Before elimination is possible, they must

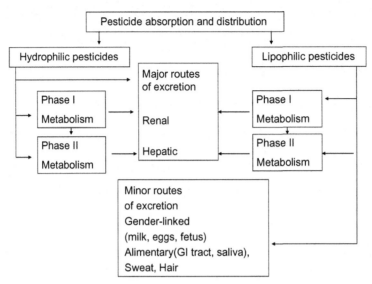

Figure 8.1 Routes of excretion in vertebrates.

first be metabolized into a form simulating that used by the body for the elimination of endogenous compounds. In general, they are metabolized by phase I and phase II xenobiotic-metabolizing enzymes to conjugation products that are more polar and hence more hydrophilic than the parent compound, and then excreted primarily by either the renal or the hepatic route. Although similar anion and cation transport systems are found in both the kidney and liver, they differ in the type of excretory products that are eliminated. The renal system eliminates molecules of molecular mass smaller than 400–500 Daltons, whereas the liver handles larger molecules. The molecular mass threshold between renal and biliary excretion varies with species (Hirom et al., 1972), although in most species there are excretory products that are excreted by both systems. In addition to excretion via the bile, highly lipophilic chemicals that are recalcitrant to metabolism may be excreted as the parent chemical by a number of alternative routes, although in terms of the overall excretion of pesticides these are generally of minor importance compared to urine and bile. General aspects of the excretion of toxicants and their metabolites may be found in Matthews (1994), Pritchard and James (1982), Tarloff and Wallace (2008, 2010) and LeBlanc (2010) and a summary of the overall process is shown in Figure 8.1.

The excretion of pesticides and their metabolites has not been extensively investigated, perhaps because the rate of excretion seldom appears to be a rate-limiting step in the ultimate expression of toxicity. However, urinary metabolites continue to be utilized as biomarkers of exposure.

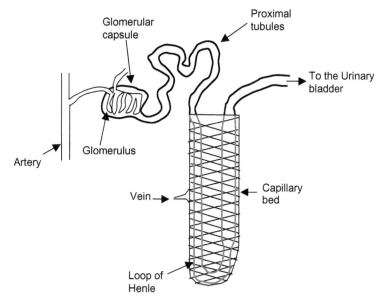

Figure 8.2 The nephron of the mammalian kidney. Reproduced from LeBlanc (2010), Figure 9.1. *Reprinted with permission of John Wiley & Sons.*

RENAL FUNCTION

Overall Aspects

The kidneys are primarily organs of excretion, and elimination by the kidney accounts for most by-products of normal body metabolism. They are also the primary organs for the excretion of polar xenobiotics and polar metabolites of lipophilic xenobiotics. Useful descriptions of kidney structure and function have recently been published by Tarloff and Wallace (2008, 2010) and LeBlanc (2010). The functional unit of the kidney, the nephron, is shown in Figure 8.2.

Glomerular Filtration

Passive filtration of the blood plasma in the glomerulus, under the influence of the blood pressure generated by the heart, is the initial step in urine formation. All molecules small enough to pass through the glomerular pores (70–100 Å) appear in the ultrafiltrate; any molecule larger than these pores or bound to molecules larger than these pores will not appear in the ultrafiltrate.

Tubular Reabsorption

Tubular reabsorption is the second major step in urine formation. Most of the reabsorption of solutes necessary for normal body function, such as amino acids, glucose, and salts, takes place in the proximal part of the tubule. This reabsorption may be active, as in the case of glucose, amino acids, and peptides, whereas water, chloride, and other ions are passively reabsorbed. Reabsorption of water and ions also occurs in the distal tubule and in the collecting duct.

Reabsorption of xenobiotics is usually passive and controlled by the same principles that regulate their passage across any membrane. That is, lipophilic compounds cross cell membranes more rapidly than polar compounds; hence, lipophilic toxicants will tend to be passively reabsorbed more than polar toxicants and, overall, elimination of polar toxicants and their polar metabolites will be facilitated.

Tubular Secretion

Tubular secretion is another important mechanism for excretion of solutes by the kidney. Secretion across the wall of the tubule is generally active, using two systems, one for the secretion of organic acids, including conjugates, and the other for the secretion of organic bases. Passive secretion may occur as a result of a process known as diffusion trapping. Un-ionized weak acids and bases pass across the membrane into the lumen of the tubule and, depending on the pH of the urine, one or the other may become ionized and unable to diffuse back across the lumen wall. Diffusion trapping is, of course, extremely sensitive to variations in urine pH, a factor that may be utilized to speed elimination of toxicants. For example, alkalinization of the urine by ingestion of bicarbonate speeds up the elimination of salicylate.

Tubular secretion, and hence excretion, of organic anions has been known to be of importance in the excretion of certain pesticides for some time (Pritchard and James, 1982). 2,4-Dichlorophenoxyacetic acid (2,4-D) and 2,4,5-trichlorophenoxyacetic acid are usually applied as salts or esters, the latter being readily hydrolyzed in the body, and studies of their excretion have emphasized the parent acids, although various conjugates are also transported by the organic anion transport system (Erne, 1966; Pritchard and James, 1982). Active tubular secretion of 2,4-D has been demonstrated in a number of species, including the rabbit (Dybing and Kolberg, 1967), rat (Fang et al., 1973), chicken (Erne and Sperber, 1974), dog (Hook et al., 1976), goat (Orberg, 1980), and flounder (Pritchard and James, 1979).

1,1,1-Trichloro-2,2-bis(p-chlorophenyl)ethane (DDT) and its principal metabolite, 1,1-dichloro-2,2-bis(p-chlorophenyl)ethylene (DDE), are highly lipophilic and the latter is recalcitrant to further metabolism. Thus, DDT and, to a greater extent, DDE are sequestered in body lipids and have an extremely long half-life in the body. Some portion of DDT, however, is metabolized to an organic acid, 2,2-bis(p-chlorophenyl) acetic acid (DDA), by dechlorination and oxidation at the one-position (Pinto et al., 1965). DDA is a substrate for the organic acid transport system (Pritchard, 1976, 1978) and, as a consequence, is excreted considerably more rapidly than DDT or DDE.

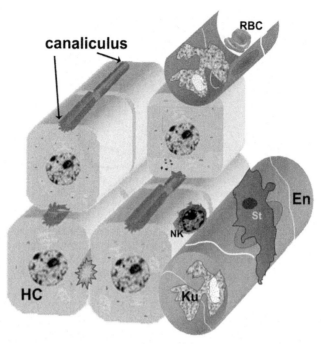

Figure 8.3 Mammalian liver structure. Reproduced from Wallace and Meyer, (2010), Figure 13.1. *Reprinted with permission of John Wiley & Sons.*

BILIARY EXCRETION

Excretion by the liver, through the biliary system, has been known for a considerable time but, because of the difficulty in obtaining uncontaminated bile, has been less intensively investigated than renal excretion. A brief review of hepatic excretion may be found in LeBlanc (2010), and a representation of liver architecture is shown in Figure 8.3. Bile is secreted by the liver cells into the bile canaliculi. It then flows into the terminal branches of the bile duct, the hepatic duct, and the gallbladder. The contents of the gallbladder are discharged into the gut under the influence of hormones whose release is triggered by food ingestion. In species that lack a gallbladder, such as the rat, bile flows continuously into the duodenum. Secretion of xenobiotics or their metabolites into the bile is largely a function of molecular mass and may occur by passive diffusion or by active transport.

Enterohepatic circulation is an important aspect of biliary excretion (Figure 8.4). Nonpolar xenobiotics are normally oxidized and then conjugated. If the molecular mass of the conjugate is appropriate for biliary excretion, it enters the gut, where hydrolysis by intestinal microflora or gut conditions may occur. The compound, then being again in a less polar form, can be reabsorbed by the intestine and returned to the

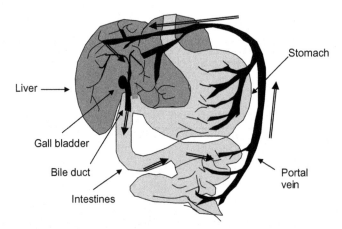

Figure 8.4 Enterohepatic circulation (as indicated by →). Polar xenobiotic conjugates are secreted into the intestine via the bile duct and gall bladder. Conjugates are hydrolyzed in the intestines, and released xenobiotics are reabsorbed and transported back to the liver via the portal vein. Reproduced from LeBlanc (2010), Figure 9.4. *Reprinted with permission of John Wiley & Sons.*

liver through portal circulation, and the process is repeated. Enterohepatic circulation thus increases the biological half-life and possibly adverse effects of toxicants, particularly to the liver. For therapeutic purposes, the cycle can be interrupted by feeding an agent that binds the hydrolysis product and prevents its reabsorption, as in the use of cholestyramine in chlordecone poisoning.

RESPIRATORY EXCRETION

Volatile toxicants such as ethanol or pesticidal fumigants may be eliminated via the lungs, as may volatile metabolites, including acetone and carbon dioxide. Respiratory excretion is not known to be an important route for excretion of pesticides, in general, or their metabolites.

OTHER ROUTES OF EXCRETION

There are a number of other, less important routes of excretion, including gender-linked routes and alimentary elimination, and several routes based on natural secretory or growth processes are known.

Gender-Linked Routes of Excretion

Certain routes of xenobiotic elimination are restricted to females, including excretion through milk, eggs, and fetus. Although such excretion is probably of minimal benefit to the mother, it may have serious consequences to the offspring.

Milk

Because milk is an emulsion of lipids in an aqueous protein solution, it may contain xenobiotics of many different physicochemical properties, ranging from polar compounds such as alcohol and caffeine, to less polar clinical drugs and to highly lipophilic chemicals such as DDT and DDE. Elimination of toxicants in milk is highly dependent on the biological half-life of the toxicant. Milk normally plays a minor role in the excretion of chemicals with short half-lives but may be important for some chemicals with long half-lives. In experimental studies with chlorinated insecticides, up to 25% of the dose administered to cows was eliminated in the milk. In some South American countries, the DDT content of human mothers' milk is close to the acceptable daily intake recommended by the World Health Organization. Although adverse effects on infants were not seen in these cases, when nursing mothers were accidentally exposed to hexachlorophene (Turkey) or polychlorinated biphenyls (PCBs) (Japan), signs of intoxication were seen in a number of infants.

Eggs

Polar toxicants and metabolites may be eliminated in egg white and lipophilic compounds in the yolk. The effects of this on developing birds are controversial but may be significant, particularly if bioconcentration has occurred in the food chain. Effects of toxicants excreted into avian eggs should not be mistaken for the well-documented eggshell thinning, which is an effect on the female reproductive system.

Fetus

The elimination of maternally derived toxicants in the fetus is of little or no benefit to the mother and, because of the generally small amounts involved, is usually of little or no harm to the fetus. However, as shown by the toxic effects of mercury, thalidomide, and diethylstilbestrol, this is not always the case.

Alimentary Elimination

Passive elimination of lipophilic toxicants directly through the wall of the alimentary canal is probably, in most cases, unimportant, at least from a quantitative viewpoint. However, although slow, it may be an important route for excretion of chlordecone, particularly if reabsorption is prevented by administration of cholestyramine.

Obscure Routes of Excretion

Because passive diffusion of lipophilic toxicants may occur across any cell membrane, it might be expected that such chemicals will appear in many body secretions, such as sweat, or in growth products, such as hair, nails, and skin. The sebaceous glands secrete an oily secretion and, probably for this reason, insecticides and PCBs have been found in human hair. Arsenic, mercury, and selenium have also been associated with hair.

Although such routes of excretion are probably only a small proportion of the total excretion of any particular xenobiotic, they may provide a noninvasive method of estimating exposure or total body burden. Analysis of bird feathers is useful for the assessment of heavy-metal exposure, and the amount of cotinine, a major metabolite of nicotine, in saliva has been used extensively as a biomarker for nicotine uptake. The excretion of atrazine in saliva has also been tested in rats as a potential biomarker of exposure in exposed workers (Lu et al., 1997).

CELLULAR ELIMINATION

To prevent concentration at toxic levels, hepatocytes and other cells have active transport processes to eliminate xenobiotics. Because the metabolism of xenobiotics generally yields products that are more polar and, consequently, have reduced capacity for passive diffusion compared to the parent compound, such transport processes are essential for cell viability. Since the publication of the second edition of the *Handbook of Pesticide Toxicology* (Krieger, 2001) there has been a dramatic increase in research on the role of transport proteins, or transporters, and, as a result, in our knowledge of their importance in the bioprocessing of xenobiotics (Miller, 2008). As yet this knowledge has not been applied extensively to pesticides; however, some general outlines of the role of transporters in pesticide bioprocessing are beginning to emerge. Many active-transport proteins have been sequenced and classified into superfamilies and families. In mammals, those transporters with xenobiotic-transport functions fall within the ABC superfamily, a group of ATP-dependent proteins (Miller, 2008). Of particular interest with regard to pesticides are *p*-glycoprotein and MRP (multidrug resistance-associated protein). MRP has the capacity to transport glutathione, sulfate, or glucuronide conjugates, whereas *p*-glycoprotein is known to transport a wide array of xenobiotics, including some pesticides.

Studies of the role of transporters in pesticide bioprocessing are generally one of two types. In the first type, the ability of the pesticide to inhibit the efflux of a chemical known to be transported by the transporter in question is measured, while in the second type the ability of the pesticide to bind to a particular transporter is measured. A useful summary of pesticides and transporters is included in Leslie et al. (2005).

The importance of the role of transporters with regard to pesticides was first illustrated by the fact that *p*-glycoprotein knockout mice died when treated with the miticide ivermectin, subsequently shown to be due to the accumulation of ivermectin in the brain because of the absence of *p*-glycoprotein in the blood-brain barrier (Schinkel et al., 1994). This and subsequent studies (Lanning et al., 1996a,b; Macdonald and Gledhill, 2007) led to the conclusion that *p*-glycoprotein provided protection from the toxic effects of both ivermectin and chlorpyrifos. Other pesticides, including metolachlor (Leslie et al., 2001), fenitrothion, methoxychlor, and chlorpropham

(Tribull et al., 2003), bind to human MRP1, and Mrp1 knockout mice are more sensitive to the toxic effects of methoxychlor. Based on studies of a p-glycoprotein polymorphism, it has been suggested that p-glycoprotein plays a protective role in Parkinson's disease (Drozdzik et al., 2003).

On the basis of their ability to inhibit the p-glycoprotein-mediated efflux of doxorubicin, several pesticides of different chemical classes were shown to bind to human p-glycoprotein (Bain and LeBlanc, 1996). The most effective were the organochlorines chlordecone, endosulfan, heptachlor and heptachlor epoxide, and the organophosphorus insecticides chlorpyrifos, chlorthiophos, dicapthon, leptophos, parathion, and phenamiphos, as well as clotrimazole and ivermectin. None of the carbamates or pyrethroids tested were effective. Lipophilicity and molecular mass were the major determinations of pesticide binding, with log K_{ow} values of 3.6–4.5 and molecular masses of 391–490 Daltons being optimal. The authors pointed out that the ability to inhibit p-glycoprotein function does not necessarily mean that the chemical will be transported. Only endosulfan, the compound with the best binding characteristics, could be shown to be transported by p-glycogen. A study of four insecticides (Sreeramula et al., 2007), methylparathion, endosulfan, cypermethrin, and fenvalerate, demonstrated that all four stimulated p-glycoprotein ATPase activity at low concentrations. At higher concentrations the stimulation was lower or, in the case of methylparathion, inhibitory. It was further demonstrated that all of these insecticides inhibited the transport of a known ligand for p-glycoprotein, tetramethylrosamine. Other recent studies have shown that several pesticides were able to inhibit the uptake of a model p-glycoprotein substrate, calcein acetoxymethyl ester, into NIH 3T3 mouse fibroblasts stably transfected with the human MDR1 gene (Pivcevic and Zaja, 2006). Of the 14 pesticides tested, endosulfan, phosalone, and propioconzole were the most active. Similarly, rotenone, diazinon, and atrazine inhibited the efflux of taxol from the basolateral to the apical side of Caco-2 cells, and rotenone and diazinon inhibited estradiol-17β-glucuronide uptake into MRP2-expressed membrane vesicles, while rotenone was a potent inhibitor of estradiol sulfate uptake into BCRP-expressing membrane vesicles (Pulsakar et al., 2006).

The possibility that p-glycoprotein is related to insect resistance to insecticides (Buss et al., 2002; Buss and Callaghan, 2008; Lanning et al., 1996a,b) has not been fully explored and, although probable, remains hypothetical.

EXCRETION OF PESTICIDES AND THEIR METABOLITES AS BIOMARKERS OF EXPOSURE

There have been a number of studies using urinary pesticides or their metabolites as biomarkers of exposures. The early studies in this area were summarized in 1989 (Wang et al., 1989), and since then some of the associated problems have been discussed (Fenske et al., 2005; Scher et al., 2007; Faustman et al., 2006; Harris et al., 2010; Naeher et al., 2010). Some of these studies involve single compounds, primarily but not exclusively organophosphorus compounds, and examples are given in Table 8.1. Other studies are surveys of populations either exposed or potentially exposed to multiple pesticides. For example, urine samples from 1000 residents of the United States were analyzed for 12 analytes potentially derived from pesticides, and 6 were frequently found (Hill et al., 1995). These, with possible parent compounds, were 2,5-dichlorophenol (from 1,4-dichlorobenzene), 2,4-dichlorophenol (from bifenox, clomethoxyfen, dichlofen-thion, etc.), 1-naphthol (from naphthalene, carbaryl, etc.), 2-naphthol (from naphthalene, etc.), 3,5,6-trichloro-2-pyridinol (from chlorpyrifos, chlorpyrifos-methyl), and pentachlorophenol (from pentachlorophenol, pentachloronitrobenzene). In another large study of multiple exposure, in tree nursery workers, only a small number, 42 of 3134, of urine samples were positive, in this case for benomyl, bifenox, and carbaryl (Lavy et al., 1993). A summary of all methods, including measurement of urinary metabolites, of estimating exposure by biomarkers was published in 2000 (Maroni et al., 2000), and appropriate analytical methods continue to be developed (e.g., for organophosphorus pesticides (OPs)) (De Alwis et al., 2008). A more recent study (McKone et al., 2007) used OP biomarker data to develop insights into the importance of various exposure sources in a cohort of almost 600. Gosselin et al. (2005) carried out a toxicokinetic modeling study of parathion and its metabolites (p-nitrophenol and alkyl phosphates) in humans to facilitate their use in exposure studies. Toxicokinetic modeling studies have also been carried out for chlorpyrifos and 2,4-D (Scher et al., 2008). Studies of exposure of schoolchildren to and the excretion of pentachlorophenol (Wilson et al., 2007) and *cis*- and *trans*-permethrin (Morgan et al., 2007; Lu et al., 2009) showed good correlation in the case of permethrin, but an excess of excretion over estimated exposure in the case of pentachlorophenol indicated that the use of urinary biomarkers is not without problems and may need to be refined for future studies, depending upon the pesticide in question.

Urinary mercapturic acids have been extensively explored for use as biomarkers of exposure, and several detailed reviews are available (De Rooij et al., 1998; Van Welie et al., 1992). The emphasis has been on industrial and environmental chemicals and this potentially valuable technique has not yet been applied extensively to pesticides. However, the soil nematocide dichloropropene was included. Methods for the detection of pyrethroid insecticides, including pyrethrins, are being developed (Leng and Greis, 2005) and validated (Barr et al., 2007), and the use of biomarkers for pesticides other than OPs is being expanded.

The effects of pesticides on the excretion of metabolites of endogenous metabolism have been explored to some extent. For example, in rats, treatment with dimethoate decreased the excretion of proline and lysine derivatives known to be collagen metabolites (Reddy et al., 1991). The mechanism of this effect was not

Table 8.1 Some Examples of the Use of Urinary Metabolites of Pesticides as Biomarkers of Exposure

Pesticide	Urinary metabolite	Reference
Acephate	Methamidophos	Bouchard et al. (2006)
Carbaryl	1-Naphthol	Meeker et al. (2005, 2007)
Chlorpyrifos	Diethylphosphate, diethylthiophosphate	Bouchard et al. (2005); Griffin et al. (1999)
Chlorpyrifos	3,5,6-Trichloro-2-pyrimidinol	Bouchard et al. (2005); Meeker et al. (2005); Garabrant et al. (2009); Whyatt et al. (2009)
Chlorpyrifos/quinalphos	Diethylphosphate, diethylthiophosphate	Vasilic et al. (1992)
Chlorpyrifos-methyl	3,5,6-Trichloro-2-pyrimidinol alkylphosphates	Aprea et al. (1997)
Deltamethrin	3-Phenoxybenzoic acid, *cis*-3-(2,2-dimethylcyclopropane-1-carboxylic acid	Ortiz-Perez et al. (2005)
Diazinon	1-Isopropyl-4-methyl-6-hydroxypyrimidine	Bouchard et al. (2006)
2,4-Dichlorophenoxyacetic acid (2,4-D)	2,4-D	Harris et al. (1992)
Dicofol	4,4′-Dichlorobenzilic acid	Nigg et al. (1991)
Guthion	Dimethylphosphorothioic acid	Franklin et al. (1981)
Malathion/thiometon	Dimethylphosphate	Vasilic et al. (1999)
Malathion/thiometon	Dimethylphosphorothioate	
Malathion/thiometon	Dimethylphosphorodithioate	
Malathion/thiometon	Mono/dicarboxylic acids	Bouchard et al. (2006); Bradman et al. (2009)
Naphthalene	1- and 2-Napthol	Meeker et al. (2007)
Organophosphorus insecticides	Alkyl phosphates	Azaroff (1999)
Permethrin (cis and trans)	3-Phenoxybenzoic acid	Morgan et al. (2007); Lu et al. (2009)

investigated and its magnitude did not seem to be large enough for practical application. N-acetylglucosamidase was found to be slightly increased in the urine of applicators exposed to the soil nematocide 1,3-dichloropropene, along with the principal metabolite of this nematocide, N-acetyl-S-(*cis*-3-chloroprop-2-enyl) cysteine (Osterloh and Feldman, 1993).

CONCLUSIONS

The excretion of pesticides in vertebrates, particularly mammals, was discussed. Renal and liver function, as they relate to excretion, were summarized, as well as the less important routes for excretion of pesticides: respiratory and alimentary. More obscure routes of excretion, including gender-linked routes such as milk, eggs, placenta, and fetus, as well as hair, sweat, etc., were briefly mentioned. The role of transporters in cellular elimination was considered, and, finally, the use of excreted pesticides and their metabolites as biomarkers of exposure was summarized.

REFERENCES

Aprea, C., Sciarra, G., Sartorelli, P., Sartorelli, E., Strambi, F., Farina, G. A., et al. (1997). Biological monitoring of exposure to chlorpyrifos-methyl by assay of urinary alkylphosphates and 3,5,6-trichloropyridinol. *J. Toxicol. Environ. Health, 50,* 581–594.

Azaroff, L. S. (1999). Biomarkers of exposure to organophosphorus insecticides among farmers families in rural El Salvador: Factors associated with exposure. *Environ. Res, 80,* 138–147.

Bain, L. J., & LeBlanc, G. A. (1996). Interaction of structurally diverse pesticides with the human MDR1 gene product p-glycoprotein. *Toxicol. Appl. Pharmacol., 141,* 288–298.

Barr, D. B., Leng, G., Berger-Preiss, E., Hoppe, H.-W., Weerasekera, G., Greis, W., et al. (2007). Cross validation of multiple methods for measuring pyrethroid and pyrethrum insecticide metabolites in human urine. *Anal. Bioanal. Chem., 389,* 811–818.

Bouchard, M., Carrier, G., Brunet, R. C., Bonvalot, Y., & Gosselin, N. H. (2005). Determination of biological reference values for chlorpyrifos metabolites in human urine using a toxicokinetic approach. *J. Occup. Environ. Hyg., 2,* 155–168.

Bouchard, M., Carrier, G., Brunet, R. C., Dumas, P., & Noisel, N. (2006). Biological monitoring of exposure to organophosphorus insecticides in a group of horticultural greenhouse workers. *Ann. Occup. Hyg., 50,* 505–515.

Bradman, A., Salvatore, A. L., Boeringer, M., Castorina, R., Snyder, J., Barr, D. B., et al. (2009). Community-based intervention to reduce pesticide exposure to farmworkers and potential take-home exposure to their families. *J. Expo. Sci. Environ. Epidemiol., 19,* 79–89.

Buss, D. S., & Callaghan, A. (2008). Interaction of pesticides with p-glycoprotein and other ABC proteins: A survey of the possible importance to insecticide, herbicide and fungicide resistance. *Pestic. Biochem. Physiol., 90,* 141–153.

Buss, D. S., McCaffery, A. R., & Callaghan, A. (2002). Evidence for p-glycoprotein modification of insecticide toxicity in mosquitoes of the Culex pipiens complex. *Med. Vet. Entomol., 16,* 218–222.

De Alwis, G. K. H., Needham, L. L., & Barr, D. B. (2008). Determination of dialkyl phosphate metabolites or organophosphorus pesticides in human urine by automated solid-phase extraction, derivatization and gas chromatography–mass spectrometry. *J. Anal. Toxicol., 32,* 721–727.

De Rooij, B. M., Commandeur, J. N. M., & Vermeulen, N. P. E. (1998). Mercapturic acids as biomarkers of exposure to electrophilic chemicals: Applications to environmental and industrial chemicals. *Biomarkers, 3,* 239–303.

Drozdzik, M., Bialecka, M., Mysliwiec, K., Honczarenko, K., Stankiewicz, J., & Sych, Z. (2003). Polymorphism in the p-glycoprotein drug transporter MDR1 gene: A possible link between environmental and genetic factors in Parkinson's disease. *Pharmacogenetics, 13,* 259–263.

Dybing, F., & Kolberg, A. (1967). Inhibition of the renal tubular transport of p-aminohippurate (Tm-PAH) in the rabbit caused by subtoxic doses of dichlorophenoxyacetate (2,4-D). *Acta Pharmacol. Toxicol., 25,* 51–61.

Erne, K. (1966). Distribution and elimination of phenoxyacetic acids in animals. *Acta Vet. Scand., 7,* 240–256.

Erne, K., & Sperber, I. (1974). Renal tubular transfer of phenoxyacetic acids in the chicken. *Acta Pharmacol. Toxicol., 35,* 233–241.

Fang, S. C., Fallin, E., Montgomery, M. L., & Freed, V. H. (1973). The metabolism and distribution of 2,4,5-trichlorophenoxyacetic acid in female rats. *Toxicol. Appl. Pharmacol., 24,* 555–563.

Faustman, E. M., Griffin, W. C., & Vigoren, E. (2006). Lessons learned from longitudinal studies to identify high exposure groups using OP pesticide urinary biomarkers. *Toxicol. Lett., 164,* S161.

Fenske, R. A., Bradman, A., Whyatt, R. M., Wolff, M. S., & Barr, D. B. (2005). Lessons learned for the assessment of children's pesticide exposure: Critical sampling and analytical issues for future studies. *Environ. Health Perspect., 113*, 1455–1462.

Franklin, C. A., Fenske, R. A., Greenhalgh, R., Mathieu, L., Denley, H. V., Leffingwell, J. T., et al. (1981). Correlation of urinary pesticide metabolite excretion with estimated dermal contact in the course of occupational exposure to guthion. *J. Toxicol. Environ. Health, 7*, 715–731.

Garabrant, D. H., Aylward, L. L., Berent, S., Chen, Q., Timchalk, C., Burns, C. J., et al. (2009). Cholinesterase inhibition in chlorpyrifos workers: Characterization of biomarkers of exposure and response in relation to urinary TCPy. *J. Expo. Sci. Environ. Epidemiol., 19*, 634–642.

Gosselin, N. H., Bouchard, M., Brunet, R. C., Dumoulin, M. J., & Carrier, G. (2005). Toxicokinetic modelling of parathion and its metabolites in humans for the determination of biological reference values. *Toxicol. Mech. Methods, 15*, 33–52.

Griffin, P., Mason, H., Heywood, K., & Cocker, J. (1999). Oral and dermal absorption of chlorpyrifos: A human volunteer study. *Occup. Environ. Med., 56*, 10–13.

Harris, S. A., Solomon, K. R., & Stephenson, G. R. (1992). Exposure of homeowners and bystanders to 2,4-dichlorophenoxyacetic acid (2,4-D). *J. Environ. Sci. Health B, 27*, 23–38.

Harris, S. A., Villeeneuve, P. J., Crawley, C. D., Mays, J. E., Yeary, R. A., Hurto, K. A., et al. (2010). National study of exposure to pesticides among professional applicators: An investigation based on urinary biomarkers. *J. Agric. Food Chem., 58*, 10253–10261.

Hill, R. H., Jr., Head, S. L., Baker, S., Gregg, M., Shealy, D. B., Bailey, S. L., et al. (1995). Pesticide residues in urine of adults living in the United States: Reference range concentrations. *Environ. Res., 71*, 99–108.

Hirom, P. C., Milburn, P., Smith, R. L., & Williams, R. T. (1972). Species variations in the threshold molecular-weight factor for the biliary excretion of organic acids. *Biochem. J., 129*, 1071–1077.

Hook, J. B., Cardona, R., Osborn, J. L., Bailie, M. D., & Gehring, P. J. (1976). The renal handling of 2,4,5-trichlorophenoxyacetic acid (2,4,5-T) in the dog. *Food Cosmet. Toxicol., 14*, 19–23.

Krieger, R. I. (Ed.). (2001). *Handbook of pesticide toxicology* (2nd ed.). San Diego, CA: Academic Press.

Krieger, R. I. (Ed.). (2010). *Handbook of pesticide toxicology* (3rd ed.). San Diego, CA: Elsevier.

Lanning, C. L., Fine, R. L., Corcoran, J. J., Ayad, H. M., Rose, R. L., & Abou-Donia, M. B. (1996). Tobacco budworm p-glycoprotein: Biochemical characterization and its involvement in pesticide resistance. *Biochim. Biophys. Acta, 1291*, 155–162.

Lanning, C. L., Fine, R. L., Sachs, C. W., Rao, U. S., Corcoran, J. J., & Abou-Donia, M. B. (1996). Chlorpyrifos oxon interacts with the mammalian multidrug resistance protein p-glycoprotein. *J. Toxicol. Environ. Health, 47*, 395–407.

Lavy, T. L., Mattice, J. D., Massey, J. H., & Skulman, B. W. (1993). Measurements of year-long exposure to tree nursery workers using multiple pesticides. *Arch. Environ. Contam. Toxicol., 24*, 123–144.

LeBlanc, G. A. (2010). Elimination of toxicants. In E. Hodgson (Ed.), *A textbook of modern toxicology* (2nd ed.). Hoboken, NJ: John Wiley & Sons.

Leng, G., & Greis, W. (2005). Simultaneous determination of pyrethroid and pyrethrin metabolites in human urine by gas chromatography–high resolution mass spectrometry. *J. Chromatogr. B Anal. Tech. Biomed. Life Sci., 814*, 285–294.

Leslie, E. M., Deeley, R. G., & Cole, S. P. (2001). Toxicological relevance of the multidrug resistance protein 1, MRP1 (ABCC1) and related transporters. *Toxicology, 167*, 3–23.

Leslie, E. M., Deeley, R. G., & Cole, S. P. C. (2005). Multidrug resistance proteins: Role of p-glycoprotein, MRP1, MRP2, and BCRP (ABCG2) in tissue defense. *Toxicol. Appl. Pharmacol., 204*, 216–237.

Lu, C., Anderson, L. C., & Fenske, R. A. (1997). Determination of atrazine levels in whole saliva and plasma in rats: Potential of salivary monitoring for occupational exposure. *J. Toxicol. Environ. Health, 50*, 101–111.

Lu, C. S., Barr, D. B., Pearson, M. A., Walker, L. A., & Bravo, R. (2009). The attribution of urban and suburban children's exposure to synthetic pyrethroid insecticides: A longitudinal study. *J. Expo. Sci. Environ. Epidemiol., 19*, 69–78.

Macdonald, N., & Gledhill, A. (2007). Potential impact of ABC1 (*p*-glycoprotein) polymorphisms on avermectin toxicity in humans. *Arch. Toxicol., 81*, 553–563.

Maroni, M., Colosio, C., Ferioli, A., & Fait, A. (2000). Introduction. *Toxicology, 143*, 5–8; Organophosphorus pesticides. *Toxicology, 143*, 9–37.

Matthews, H. B. (1994). Excretion and elimination of toxicants and their metabolites. In E. Hodgson & P. E. Levi (Eds.), *Introduction to biochemical toxicology* (2nd ed.). East Norwalk, CT: Appleton & Lange (Chap. 8).

McKone, T. E., Castorina, R., Harnly, M. E., Kuwabara, Y., Eskenazi, B., & Bradman, A. (2007). Merging models and biomonitoring data to characterize sources and pathways of human exposure to organophosphorus pesticides in the Salinas valley of California. *Environ. Sci. Technol., 41*, 3233–3240.

Meeker, J. D., Barr, D. B., Ryan, L., Herrick, R. F., Bennett, D. H., Bravo, R., et al. (2005). Temporal variability of urinary levels of nonpersistent pesticides in adult men. *J. Expo. Anal. Environ. Epidemiol, 15*, 271–281.

Meeker, J. D., Barr, D. B., Serdar, B., Rappaport, S. M., & Hauser, R. (2007). Utility of 1-naphthol and 2-naphthol levels to assess environmental carbaryl and naphthalene exposure in an epidemiological study. *J. Expo. Sci. Environ. Epidemiol., 17*, 314–320.

Miller, D. S. (2008). Cellular transport and elimination. In R. C. Smart & E. Hodgson (Eds.), *Molecular and biochemical toxicology*. Hoboken, NJ: John Wiley & Sons.

Morgan, M. K., Sheldon, L. S., Croghan, C. W., Jones, P. A., Chuang, J. C., & Wilson, N. K. (2007). An observational study of 127 preschool children at their homes and daycare centers in Ohio: Environmental pathways to cis- and trans-permethrin exposure. *Environ. Res., 104*, 266–274.

Naeher, L. P., Tulve, N. S., Egeghy, P. P., Bar, D. B., Adetona, O., Fortmann, R. C., et al. (2010). Organophosphorus and pyrethroid insecticide urinary concentrations in young children living in a southeastern United States city. *Sci. Total Environ., 408*, 1145–1153.

Nigg, H. N., Stamper, J. H., Deshmukh, S. N., & Queen, R. M. (1991). 4,49-Dichlorobenzilic acid urinary excretion by dicofol pesticide applicators. *Chemosphere, 22*, 365–373.

Orberg, A. (1980). Observations on 2,4-dichlorophenoxyacetic (2,4-D) excretion in the goat. *Acta Pharmacol. Toxicol., 46*, 78–80.

Ortiz-Perez, M. D., Torres-Dorsal, A., Batres, L. E., Lopez-Guzman, O. D., Grimaldo, M., & Carranza, C., et al. (2005). Environmental health assessment of deltamethrin in a malarious area of Mexico: Environmental persistence, toxicokinetics, and genotoxicity in exposed children. *Environ. Health Perspect., 113*, 782–786.

Osterloh, J. D., & Feldman, B. J. (1993). Urinary protein markers in pesticide applicators during a chlorinated hydrocarbon exposure. *Environ. Res., 63*, 171–181.

Pinto, J. D., Camien, M. N., & Dunn, M. S. (1965). Metabolic fate of p,p9-DDT [1,1,1-trichloro-2,2-bis(p-chlorophenyl) ethane] in rats. *J. Biol. Chem., 240*, 2148–2157.

Pivcevic, B., & Zaja, R. (2006). Pesticides and their binary combinations as p-glycoprotein inhibitors in NIH 3T3/MDR1 cells. *Environ. Toxicol. Pharmacol., 22*, 268–276.

Pritchard, J. B. (1976). In vitro analysis of 2,2-bis(p-chlorophenyl) acetic acid (DDA) handling by rat kidney and liver. *Toxicol. Appl. Pharmacol., 38*, 621–630.

Pritchard, J. B. (1978). Kinetic analysis of renal handling of 2,2-bis(p-chlorophenyl) acetic acid by rat. *J. Pharmacol. Exp. Ther., 205*, 9–18.

Pritchard, J. B., & James, M. O. (1979). Determinants of the renal handling of 2,4-dichlorophenoxyacetic by winter flounder. *J. Pharmacol. Exp. Ther., 208*, 208–286.

Pritchard, J. B., & James, M. O. (1982). Metabolism and urinary excretion. In W. B. Jakoby, J. R. Bend, & J. Caldwell (Eds.), *Metabolic basis of detoxiation: Metabolism of functional groups*. San Diego, CA: Academic Press.

Pulsakar, S., Williams, D. A., LeDuc, B., Liu, N., & Xia, C. (2006). Interaction of structurally diverse pesticides with multidrug resistance proteins (P-GP, MRP2 and BCRP). *Drug Metab. Rev., 38*(Suppl. 2), 243.

Reddy, P. N., Raj, G. D., & Dhar, S. C. (1991). Toxicological effects of an organophosphorus pesticide (dimethoate) on urinary collagen metabolites in normal and high protein diets fed female albino rats. *Life Sci., 49*, 1309–1318.

Scher, D. P., Alexander, B. H., Adgate, J. L., Eberly, L. E., Mandel, J. S., Acquavella, J. F., et al. (2007). Agreement of pesticide biomarkers between morning void and 24-h urine samples from farmers and their children. *J. Expo. Sci. Environ. Epidemiol., 17*, 350–357.

Scher, D. P., Sawchuck, R. J., Alexander, B. H., & Adgate, J. L. (2008). Estimating absorbed dose of pesticides in a field study using biomonitoring data and pharmacokinetic models. *J. Toxicol. Environ. Health A, 71,* 373–383.

Schinkel, A. H., Smit, J. J. M., van Tellingen, O., Beijnen, J. H., Wagenaar, E., van Deemter, L., et al. (1994). Disruption of the mouse mdr1a *p*-glycoprotein gene leads to a deficiency in the blood–brain barrier and to increased sensitivity to drugs. *Cell, 77,* 491–502.

Sreeramula, A., Liu, R., & Sharom, F. J. (2007). Interaction of insecticides with mammalian p-glycoprotein and their effect on its transport function. *Biochim. Biophys. Acta, 1768,* 1750–1757.

Tarloff, J. B., & Wallace, A. D. (2008). Biochemical mechanisms of renal toxicity. In R. C. Smart & E. Hodgson (Eds.), *Molecular and biochemical toxicology* (4th ed.). Hoboken, NJ: John Wiley & Sons.

Tarloff, J. B., & Wallace, A. D. (2010). Nephrotoxicity. In E. Hodgson (Ed.), *A textbook of modern toxicology* (4th ed). Hoboken, NJ: John Wiley & Sons.

Tribull, T. E., Bruner, R. H., & Bain, L. J. (2003). The multidrug resistance-associated protein 1 transports methoxychlor and protects the seminiferous epithelium from injury. *Toxicol. Lett., 142,* 61–70.

Van Welie, R. T. H., van Dijck, R. G. J. M., & Vermeulen, N. P. E. (1992). Mercapturic acids, protein adducts, and DNA adducts as biomarkers of electrophilic chemicals. *Crit. Rev. Toxicol., 22,* 271–306.

Vasilic, Z., Drevenkar, V., Rumenjak, V., Stengl, B., & Frobe, Z. (1992). Urinary excretion of diethylphosphorus metabolites in persons by quinalphos or chlorpyrifos. *Arch. Environ. Contam. Toxicol., 22,* 351–357.

Vasilic, Z., Stengl, B., & Drevenkar, V. (1999). Dimethylphosphorus metabolites in serum and urine of persons poisoned by malathion or thiometon. *Chem. Biol. Interact., 119–120,* 479–487.

Wallace, A. D., & Meyer, S. A. (2010). Hepatotoxicity. In E. Hodgson (Ed.), *A textbook of modern toxicology.* Hoboken, NJ: John Wiley & Sons.

Wang, R. G. M., Franklin, C. A., Honeycutt, R. C., & Reinert, J. C. (Eds.). (1989). *Biological monitoring for pesticide exposure: Measurement, estimation and risk reduction.* Washington, DC: American Chemical Society.

Whyatt, R. M., Garfinkel, R., Hoepner, L. A., Andrews, H., Holmes, D., Williaams, M. K., et al. (2009). A biomarker validation study of prenatal chlorpyrifos exposure within an inner-city cohort during pregnancy. *Environ. Health Perspect., 117,* 559–567.

Wilson, N. K., Chuang, J. C., Morgan, M. K., Lordo, R. A., & Sheldon, L. S. (2007). An observational study of the potential exposures of preschool children to pentachlorophenol, bisphenol A and nonylphenol at home and daycare. *Environ. Res., 103,* 9–20.

CHAPTER 9

Biotransformation of Individual Pesticides: Some Examples

Ernest Hodgson
North Carolina State University, Raleigh, NC, USA

OUTLINE

INTRODUCTION

This chapter is intended to show how the general features of pesticide metabolism are expressed in surrogate species and, in some cases, in humans, using individual, well-known pesticides as examples. Of particular interest is the integration of the role of several phase I and/or phase II enzymes to effect the overall metabolism of a single chemical entity. In several cases both detoxication and activation pathways are apparent.

SELECTED PESTICIDES

Carbamate Insecticides—Carbaryl

Carbaryl (chemical name 1-naphthalenyl methylcarbamate, CAS No. 63-25-2) is sold under many trade names, the most common being Sevin. It is widely used in agriculture, in horticulture, and in residential settings. The primary mechanism of action is reversible inhibition of acetylcholinesterase and it is generally regarded as being safe with respect to human health. Metabolism in the rat is shown in Figure 9.1 (Blacker et al., 2010).

Pesticide Biotransformation and Disposition
DOI: 10.1016/B978-0-12-385481-0.00009-5

195

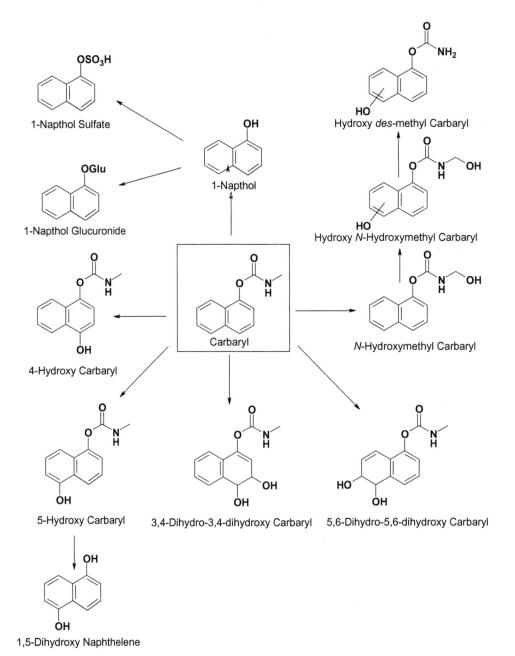

Figure 9.1 In vitro metabolism of carbaryl. *Reproduced from Blacker et al. (2010), Figure 74.1.*

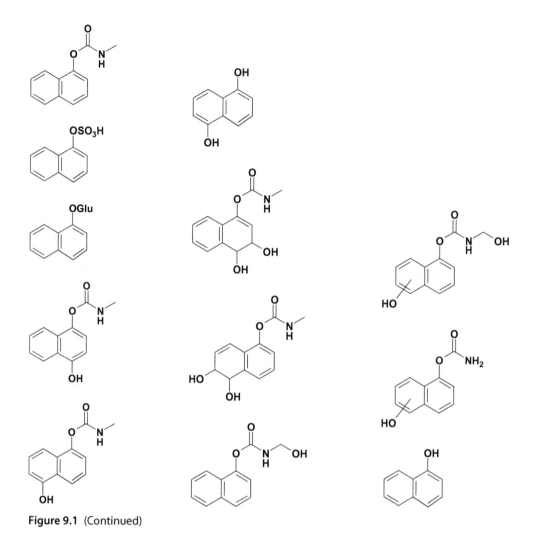

Figure 9.1 (Continued)

The principal metabolites produced in vitro by human liver microsomes are 5-hydroxycarbaryl, 4-hydroxycarbaryl, and carbaryl methylol. Most of the 16 recombinant human CYP isoforms tested had some activity toward carbaryl, with CYP1A1 and CYP1A2 having the greatest ability to form 5-hydroxycarbaryl, CYP3A4 and CYP1A1 to form 4-hydroxycarbaryl, and CYP2B6 to form carbaryl methylol (Tang et al., 2002).

Organochlorine Insecticides—DDT

Although DDT has long been banned in the United States, Europe, and elsewhere, it is still used to some extent in public health, for example in mosquito control, elsewhere

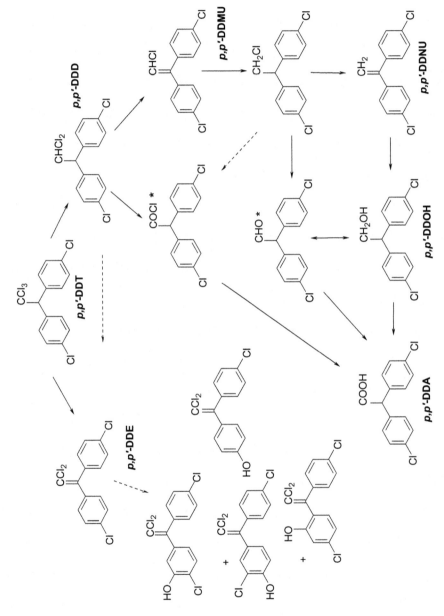

Figure 9.2 In vitro metabolism of DDT. *Reproduced from Smith (2010), Figure 93.2.*

in the world. The chemical name of DDT is 1,1′-(2,2,2-trichloroethylidene)-bis(4-chlorobenzene) (CAS No. 50-29-3). It is known almost universally as DDT or p,p-DDT. DDT and its metabolite DDE are highly persistent, and environmental and human tissue residues are still detectable, even in areas in which DDT has not been used for decades.

Metabolism of DDT in the rat is shown in Figure 9.2 and is reviewed by Smith (2010).

Organophosphorus Insecticides

There are numerous chemical types of organophosphorus insecticides, including phosphates, phosphorothioates, phosphorphorodithioates, phosphonates, etc. These and others are classified by Chambers et al. (2010) in a scheme based on the central phosphorus atom and the four atoms immediately surrounding it.

Chlorpyrifos

Chlorpyrifos is an organophosphorothionate insecticide with the chemical name O,O-diethyl-O-(3,5,6-trichloro-2-pyridinyl) phosphorothionate (CAS No. 2921-89-2). It is widely used in agriculture. In the United States and elsewhere it was also used in public health and residential pest control, but recently the U.S. Environmental Protection Agency has banned residential use in the United States.

The oxon derivative is formed metabolically by CYP isoforms and is an irreversible inhibitor of acetylcholinesterase. Chlorpyrifos is also known to have other, noncholinergic effects. These include inhibition, in humans, of the oxidative metabolism of both testosterone and estradiol.

Metabolism of chlorpyrifos in human hepatocytes is shown in Figure 9.3a and that of chlorpyrifos oxon in Figure 9.3b. A review of chlorpyrifos toxicology is provided by Testai et al. (2010).

Phorate

Phorate is a phosphorodithioate insecticide (chemical name O,O-diethyl-S-ethylthiomethylphosphorodithioate, CAS No. 298-02-2). Phorate is used in agriculture as a soil and systemic insecticide and sold under a number of names, the most common of which is Thimet. The mode of action of phorate is, first, activation to an oxon by CYP, followed by inhibition of acetylcholinesterase by the oxon. While both the $P = S$ sulfur and the thioether sulfur are metabolized by cytochrome P450, the thioether sulfur is also attacked by the flavin-containing monooxygenase (Figure 9.4) (Levi and Hodgson, 1988).

Neonicotinoid Insecticides—Imidacloprid

Imidacloprid (chemical name N-[1-((6-chloro-3-pyridyl)methyl)-4,5-dihydroimidazol-2-yl], CAS No. 138261-41-3) is an important representative of a relatively new class of

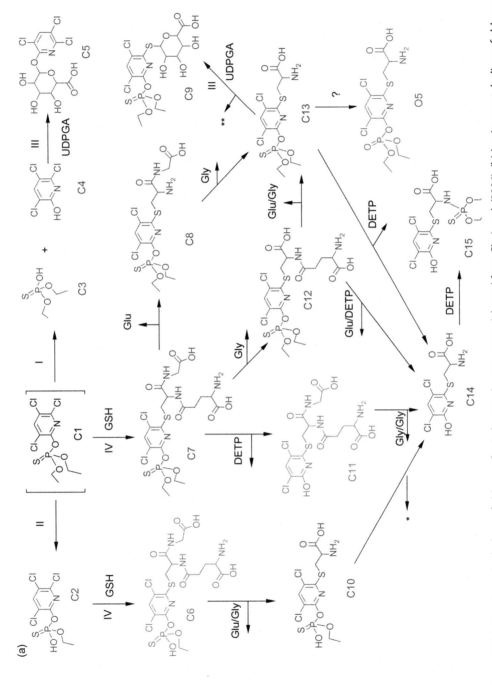

Figure 9.3 (a) In vitro metabolism of chlorpyrifos in human hepatocytes. *Adapted from Choi et al. (2006).* (b) In vitro metabolism of chlorpyrifos oxon in human hepatocytes. *Adapted from Choi et al. (2006).*

Figure 9.3 (Continued)

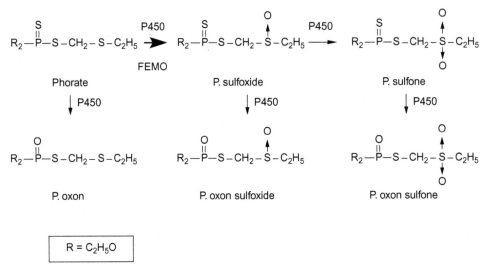

Figure 9.4 In vitro oxidative metabolism of phorate by mouse liver enzymes.

insecticide, the neonicotinoids, synthetic chemicals based on nicotine. Imidacloprid is systemic in both plants and animals; is widely used for agricultural, veterinary, and residential purposes; and may be the most widely used of all insecticides.

Its mechanism of action involves binding to the nicotinic acetylcholine receptor, and its metabolism is illustrated in Figure 9.5.

Phenylpyrazole Insecticides—Fipronil

Fipronil (chemical name 5-amino-1-[2,6-dichloro-4-(trimethylmethyl)sulfinyl]-1*H*-pyrazole-3-carbonitrile, CAS No. 120068-37-3) is a member of another relatively new class of pesticides, the phenylpyrazole insecticides. Fipronil is a widely used, broad-spectrum insecticide, with applications in crop production and in veterinary practice. Its mechanism of action involves binding to the GABA receptor. Metabolism in both surrogate animals and humans consists primarily of oxidation to the sulfone, the only metabolite produced in mice and by human liver microsomes (Figure 9.6). In the latter case CYP3A4 and CYP2C19 are responsible for essentially all of the activity (Tang et al., 2004).

Pyrethroid Insecticides—Permethrin

Permethrin (chemical name 3-(2,2-dichloroethenyl)-2,2-dimethylcyclopropanecarboxylic acid, CAS No. 52645-53-1) is representative of a group of chemicals, the synthetic pyrethroids, whose structures are based on the botanical insecticide pyrethrin.

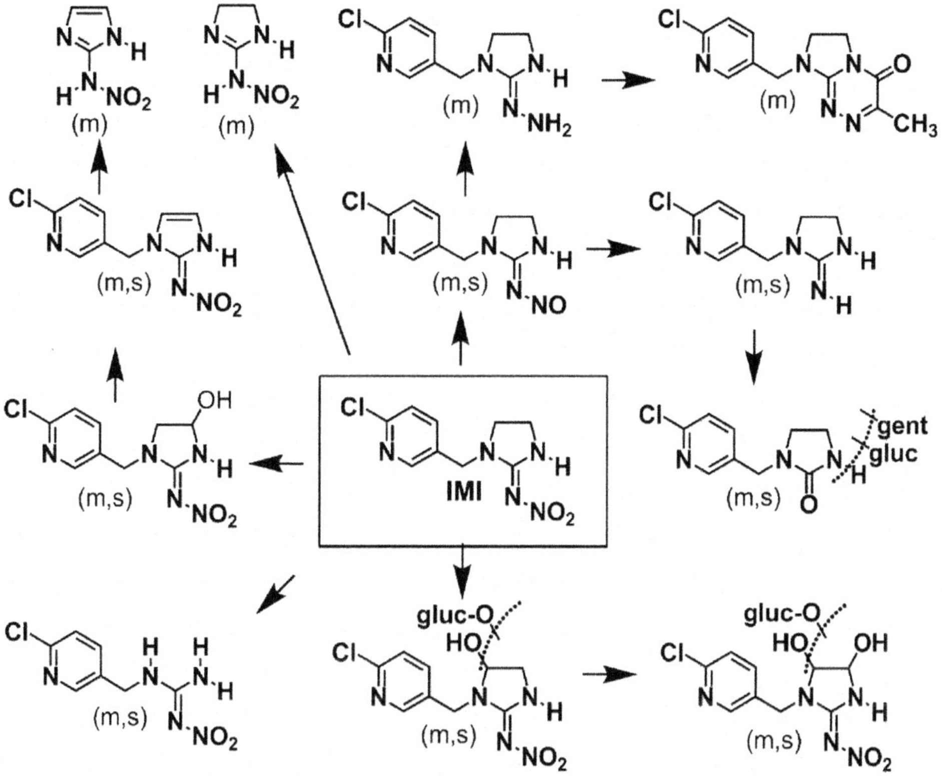

Figure 9.5 In vitro metabolism of imidacloprid in mice (m) and spinach (s). *Reproduced from Casida (2011).*

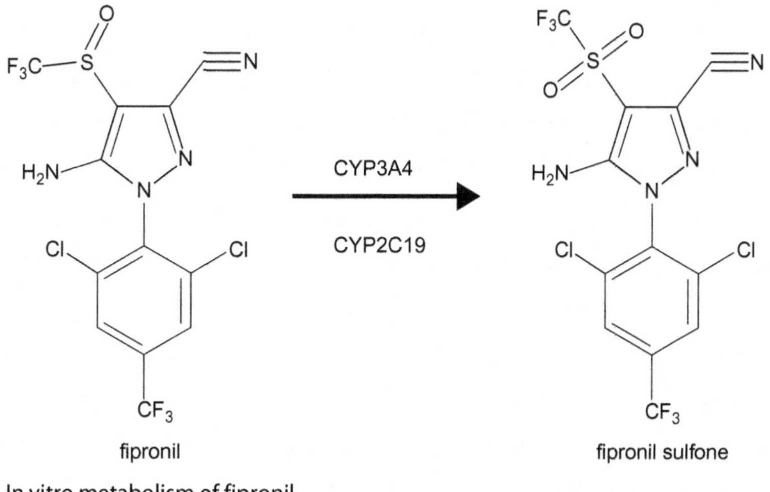

Figure 9.6 In vitro metabolism of fipronil.

Permethrin is widely used in agriculture and, although its mechanism of action, interaction with the sodium channel in nerve membranes, is similar to that of DDT, it is not persistent in vivo or in the environment. The metabolism of permethrin has been studied extensively in surrogate animals and, to a much lesser extent, in humans. A detailed, and complex metabolic pathway, along with those for other synthetic pyrethroids, is shown in Kaneko (2010). Although these metabolic pathways emphasize the involvement of cytochrome P450, the possibility has been raised that, at least in humans (Choi et al., 2003), this is, to a large extent, secondary metabolism of primary metabolites formed by hydrolysis and subsequently acted upon by the ethanol and aldehyde dehydrogenases (Figure 9.7).

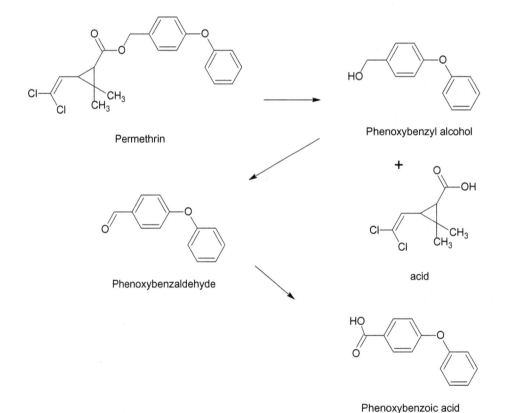

Figure 9.7 In vitro metabolism of pyrethroid insecticides—permethrin. *Reproduced from Kaneko (2010), Figure 76.19.*

Chloroacetamide Herbicides—Alachlor and Butachlor

Alachlor and butachlor are representative of the chloroacetamide class of herbicides, their chemical names and CAS numbers being as follows: alachlor, N-methoxymethyl-2′,6′-diethyl-2-chloroacetanilide and 15972-60-8; butachlor, N-(butoxyethyl)-2-chloro-2′,6′-diethylacetanilide and 23184-66-99. Alachlor, but not butachlor, is widely used in the United States, primarily on corn. The metabolism of these herbicides has been studied in surrogate animals and humans (Figure 9.8) and their toxicology summarized (Coleman et al., 2000; Heydens et al., 2010).

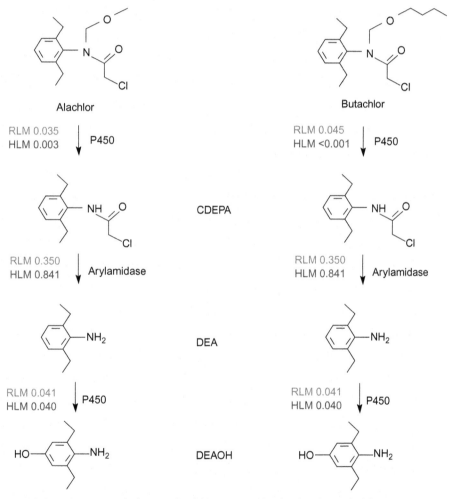

Figure 9.8 In vitro metabolism of chloroacetamide herbicides—alachlor and butachlor. Abbreviations: HLM, human liver microsomes; RLM, rat liver microsomes; CDEPA, 2-chloro-N-(2,6-diethylphenylacetamide); DEA, 2,6-diethylaniline; DEAOH, 4-hydroxy-2,6-diethylaniline.

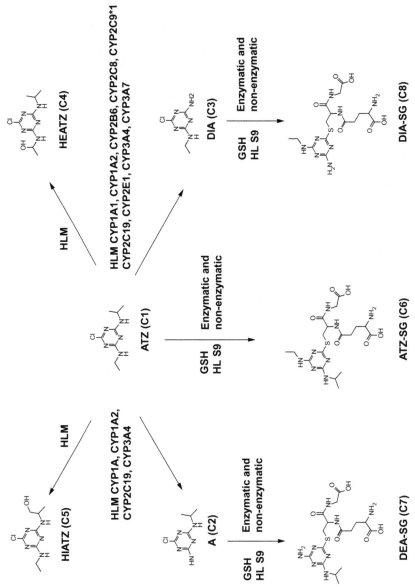

Figure 9.9 In vitro metabolism of triazine herbicides—atrazine in humans. *Adapted from Joo et al. (2010).*

Triazine Herbicides—Atrazine

Atrazine (chemical name 2-chloro-4-ethylamino-6-isopropylamino-1,3,5-triazine, CAS No. 1912-24-9) is used extensively for the control of broadleaf weeds in agricultural and roadway applications. Despite its frequent presence as a well-water contaminant, the U.S. Environmental Protection Agency concluded it was unlikely to be a carcinogenic hazard to humans. It is, however, recognized as an endocrine disruptor. The metabolism of atrazine has been investigated in surrogate animals and humans. Lang et al. (1997) concluded that CYP1A2 was primarily responsible for atrazine metabolism in human liver and, recently, Joo et al. (2010) identified further metabolites and showed that CYP1A2 and CYP3A4 were about equally responsible for atrazine phase I metabolism (Figure 9.9).

CONCLUSIONS

It is clear from these examples that many phase I and phase II xenobiotic-metabolizing enzymes (XMEs) are involved in the metabolism of pesticides and that individual XME isoforms are frequently involved in the metabolism of pesticides from different chemical and use classes. Of particular interest is the integration of the role of several phase I and/or phase II enzymes to effect the overall metabolism of a single chemical entity. The recent emphasis on human studies of pesticide metabolism is likely to develop quickly as new risk assessment paradigms dependent upon human cell lines are implemented.

ACKNOWLEDGMENTS

Several authors of figures were of great help in providing the originals of figures from the *Handbook of Pesticide Toxicology,* 3rd edition, or otherwise of assistance with the adaptation of figures. Drs. M. Krolski, A.G. Smith, and A.D. Wallace are gratefully acknowledged.

REFERENCES

Blacker, A. M., Lunchick, C., Lasserre-Bigot, D., Payraudeau, V., & Krolski, M. (2010). Toxicological profile of carbaryl. In R. Krieger (Ed.), *Handbook of pesticide toxicology* (3rd ed.). San Diego, CA: Elsevier (Chap. 74)

Casida, J. E. (2011). Neonicotinoid metabolism: Compounds, substrates, pathways, organisms and relevance. *J. Agric. Food Chem., 59,* 2923–2931.

Chambers, H. W., Meek, E. C., & Chambers, J. E. (2010). Chemistry of organophosphorus insecticides. In R. Krieger (Ed.), *Handbook of pesticide toxicology* (3rd ed.). San Diego, CA: Elsevier (Chap. 64).

Choi, J., Rose, R. L., & Hodgson, E. (2003). In vitro human metabolism of permethrin: The role of human alcohol and aldehyde dehydrogenases. *Pestic. Biochem. Physiol., 73,* 117–128.

Choi, K., Joo, H., Rose, R. L., & Hodgson, E. (2006). Metabolism of chlorpyrifos and chlorpyrifos oxon by human hepatocytes. *J. Biochem. Mol. Toxicol., 20,* 279–291.

Coleman, S., Linderman, R., Hodgson, E., & Rose, R. L. (2000). Comparative metabolism of chloroacet-amide herbicides and selected metabolites in human and rat liver microsomes. *Environ. Health Perspect.*, *108*, 1151–1157.

Heydens, W. F., Lamb, I. C., & Wilson, A. G. E. (2010). Chloroacetanilides. In R. Krieger (Ed.), *Handbook of pesticide toxicology* (3rd ed.). San Diego, CA: Elsevier (Chap. 82).

Joo, H., Choi, K., & Hodgson, E. (2010). Human metabolism of atrazine. *Pestic. Biochem. Physiol.*, *98*, 73–79.

Kaneko, K. (2010). Pyrethroid chemistry and metabolism. In R. Krieger (Ed.), *Handbook of pesticide toxicology* (3rd ed.). San Diego, CA: Elsevier (Chap. 76).

Lang, D. H., Rettie, A. E., & Bocker, R. H. (1997). Identification of the enzymes involved in the metab-olism of atrazine, ametryne and terbutryne in human liver microsomes. *Chem. Res. Toxicol.*, *10*, 1037–1044.

Levi, P. E., & Hodgson, E. (1988). Stereospecificity of the oxidation of phorate and phorate sulphoxide by purified FAD-containing monooxygenase and cytochrome P450. *Xenobiotica*, *1*, 29–39.

Smith, A. G. (2010). Toxicology of DDT and some analogues. In R. Krieger (Ed.), *Handbook of pesticide toxicology* (3rd ed.). San Diego, CA: Elsevier (Chap. 93).

Tang, J., Cao, Y., Rose, R. L., & Hodgson, E. (2002). In vitro metabolism of carbaryl by human cyto-chrome P450 and its inhibition by chlorpyrifos. *Chem. Biol. Interact.*, *141*, 229–241.

Tang, J., Usmani, K. A., Hodgson, E., & Rose, R. L. (2004). In vitro metabolism of fipronil by human and rat cytochrome P450 and its interactions with testosterone and diazepam. *Chem. Biol. Interact.*, *147*, 319–329.

Testai, E., Buratti, F. M., & Di Consiglioni, E. (2010). Chlorpyrifos. In R. Krieger (Ed.), *Handbook of pesti-cide toxicology* (3rd ed.). San Diego, CA: Elsevier (Chap. 70).

CHAPTER 10

Summary, Conclusions, and Future Developments

Ernest Hodgson
North Carolina State University, Raleigh, NC, USA

Outline

INTRODUCTION

There is a large literature on the metabolism of pesticides in both target and surrogate animal species and a much smaller, but growing, literature on human metabolism of pesticides. However, given the number of pesticides and the importance of this class of chemicals, much more work is necessary. At the same time, given the effort, time, and cost in arriving at a complete understanding of pesticide metabolism it appears unlikely that all of the needed studies can be accomplished by traditional methods. Studies of pesticide disposition, particularly biotransformation, are critical to the understanding of the toxic mode of action in both target and nontarget organisms if new selective pesticides are to be discovered. The balance between activation to toxic metabolites and detoxication is crucial to this effort.

Disposition studies are also essential for risk analysis. They make possible physio-logically based pharmacokinetic (PBPK) studies, required for regulatory and other purposes. Science-based risk assessment is possible only if the mechanism of action at the molecular level can be defined, and only if all active metabolites have been identified and their interactions at the site of action determined. Quantitative structure-activity relationship (QSAR) studies, important for the prediction of both effectiveness and toxicity, likewise depend on detailed knowledge of metabolites and their formation.

Pesticide Biotransformation and Disposition
DOI: 10.1016/B978-0-12-385481-0.00010-1

Given the ready availability of hepatocytes, cell fractions, cell lines, and recombinant xenobiotic-metabolizing enzymes (XMEs), all derived from humans, ethical human studies have been relatively easy to conduct since the beginning of the 21st century. Such studies are essential if variation within the human population is part of the risk assessment paradigm, and if subpopulations and individuals at increased risk are to be identified.

Finally, since pesticides are, more often than not, used in mixtures or in temporal proximity that is so close as to have the same implications for risk analysis as mixtures, the question of interactions between the pesticides in mixtures and between pesticides and endogenous metabolites is beginning to be addressed, at the metabolic and other levels.

CONTINUING NEED FOR PESTICIDE METABOLISM STUDIES

In view of the needs outlined above it is clear first, that metabolic studies must continue and second, that new techniques need to be brought to bear on the various problems involved. Although production of food and fiber without the use of pesticides ("green agriculture") may be a worthy goal, it seems unlikely, in the short and intermediate future, that advances in that direction can produce the food and fiber necessary to care for a world population of over 6 billion and growing.

Given the emerging changes in risk assessment (National Research Council, 2007; Hodgson, 2010a,b; Kullman et al., 2010) that rely heavily on human cell lines and the techniques of genomics, proteomics, metabolomics, QSAR, and bioinformatics, the nature of human studies will doubtless change, but their importance will increase. Examination for potential toxicity of the very large number of chemicals in general use, using high-throughput genomic, proteomic, and metabolic methods, will go far to make practical a task that currently, because of the huge number of chemicals involved, is not feasible. Experience would suggest, however, that both false positives and false negatives will occur and that detailed studies of single chemicals will still be needed. In the specific case of pesticides. Because of the regulatory requirements for registration and use, it is likely that the detailed studies of both single chemicals and the mixtures actually used in the field will continue alongside the new techniques. Detailed studies on the pesticide mixtures commonly in use and PBPK studies of one pesticide in the presence of another will be an essential starting point.

Improved human health risk analysis is vital when, although it is becoming increasingly clear that the risks to human health from pesticides are minimal to the general population when the pesticides are properly evaluated and used, there are significant risks to applicators and other users. By some statistical measures, agriculture is now the most hazardous occupation in the United States (U.S. Bureau of Labor Statistics, 2011).

NEW APPROACHES TO PESTICIDE METABOLISM

Analytical Chemistry

The current status of the analytical chemistry techniques used in pesticide metabolism studies is summarized in Chapter 2. It should be noted that the improvement in these techniques with regard to both sensitivity and chemical structure characterization has, in the past few decades, been dramatic and far exceeds our capacity to understand the significance of the minute quantities of a toxicant that can be detected and characterized. It is also true that, without these advances, it would not be possible to realize the potential of proteomics and metabolomics.

Human Studies

Although metabolic studies on surrogate animals will continue to be necessary, the current expansion of human studies will continue, particularly since the new paradigm for risk assessment (National Research Council, 2007) depends heavily on human cell lines and other human-derived information, thus minimizing the need for extrapolation from surrogate animal to human. These studies have been facilitated by the availability of human hepatocytes, other human-derived cell lines, human cell fractions such as microsomes and cytoplasm, and recombinant human XMEs. Although studies of the effects of polymorphisms on pesticide metabolism are few, new genomic techniques will greatly speed up this approach.

Toxicogenomics

Toxicogenomics describes how chemical stressors such as pesticides affect the entire genome, combining information from genome-wide mRNA profiling; protein profiling, or proteomics; changes in the metabolome, or metabolomics; genetic susceptibility; gene-environment interactions; and the computational models necessary to make generalizations from the data produced and thus relate chemical toxicity to human disease (Tennant, 2002; Waters and Fostel, 2004; Bogert, 2007).

Metabolomics

Metabolomics, as used in toxicogenomics, describes the changes in the metabolome, the total of all normal metabolites. However, the techniques of metabolomics can be used in a more restricted sense in pesticide metabolism studies, namely to describe the total of all metabolites (including all primary and secondary metabolites) of a toxicant administered to an organism or cell line.

Bioinformatics

Bioinformatics was originally defined as the application of information technology to molecular biology. While this is still of critical importance, bioinformatics has become much broader and, in addition to other aspects, now plays a major role in new approaches to the toxicology of chemicals, particularly in the development of programs for handling microarray data. While such data are probably of most importance in mode-of-action studies, the effects on the expression of XME genes are of importance in the study of pesticide metabolism. Bioinformatics is characterized by computationally intensive methodology and includes the design of large databases and the techniques needed for their manipulation, including data mining. Again, such applications to pesticides are few (e.g., Wallace et al., 2011) but they will certainly increase. A useful summary of bioinformatics is provided by Kullman et al. (2010).

CONCLUSIONS

As long as pesticides are used, and both target and nontarget species are exposed to them, the metabolism of pesticides will continue to be of importance and will continue to be investigated. It is also clear, as indicated above, that the techniques used will change as will the purposes to which the results are applied, but these will still include such activities as risk assessment and the development of selective pesticides.

REFERENCES

Bogert, C. J. (2007). Predicting interactions from mechanistic information: Can omic data validate theories? *Toxicol. Appl. Pharmacol.*, *223*, 114–120.

Hodgson, E. (2010a). Introduction to toxicology. In E. Hodgson (Ed.), *A textbook of modern toxicology* (4th ed.). Hoboken, NJ: John Wiley & Sons (Chap. 1).

Hodgson, E. (2010b). Future considerations. In E. Hodgson (Ed.), *A textbook of modern toxicology* (4th ed.). Hoboken, NJ: John Wiley & Sons (Chap. 29).

Kullman, S. W., Mattingly, C. J., Meyer, J. N., & Whitehead, A. (2010). Perspectives on informatics in toxicology. In E. Hodgson (Ed.), *A textbook of modern toxicology* (4th ed.). Hoboken, NJ: John Wiley & Sons (Chap. 28).

Committee on Toxicity Testing and Assessment of Environmental Agents, National Research Council. (2007). *Toxicity testing in the 21st century: A vision and a strategy*. Washington, DC: National Academies Press.

Tennant, R. W. (2002). The National Center for Toxicogenomics: Using new techniques to inform mechanistic toxicology. *Environ. Health Perspect*, *110*, A8–A10.

U.S. Bureau of Labor Statistics. (2011). *U.S. Department of Labor current population survey, census of fatal occupational injuries*. Washington, DC: U.S. Census Bureau.

Wallace, A. D., Shah, R., Choi, K., Joo, H., Hodgson, E. (2011). Microarray analysis of gene expression changes in human hepatocytes after chlorpyrifos exposure. Abstract 2096. Society of Toxicology Annual Meeting, March 6–10, 2011, Washington, DC.

Waters, M. D., & Fostel, J. M. (2004). Toxicogenomics and systems biology: Aims and prospects. *Nat. Rev. Genet.*, *5*, 936–948.

INDEX

Printed in the United States
By Bookmasters